Intelligent Systems Reference Library

Volume 116

About this Series

The aim of this series is to publish a Reference Library, including novel advances and developments in all aspects of Intelligent Systems in an easily accessible and well structured form. The series includes reference works, handbooks, compendia, textbooks, well-structured monographs, dictionaries, and encyclopedias. It contains well integrated knowledge and current information in the field of Intelligent Systems. The series covers the theory, applications, and design methods of Intelligent Systems. Virtually all disciplines such as engineering, computer science, avionics, business, e-commerce, environment, healthcare, physics and life science are included.

More information about this series at http://www.springer.com/series/8578

Ioannis Hatzilygeroudis · Vasile Palade
Jim Prentzas
Editors

Advances in Combining Intelligent Methods

Postproceedings of the 5th International Workshop CIMA-2015, Vietri sul Mare, Italy, November 2015 (at ICTAI 2015)

Springer

Editors
Ioannis Hatzilygeroudis
Department of Computer Engineering and Informatics, School of Engineering
University of Patras
Patras
Greece

Vasile Palade
Department of Computing, Faculty of Engineering and Computing
Coventry University
Coventry
UK

Jim Prentzas
Laboratory of Informatics, Department of Education Sciences in Early Childhood, School of Education Sciences
Democritus University of Thrace
Alexandroupoli
Greece

ISSN 1868-4394 ISSN 1868-4408 (electronic)
Intelligent Systems Reference Library
ISBN 978-3-319-83467-2 ISBN 978-3-319-46200-4 (eBook)
DOI 10.1007/978-3-319-46200-4

Printed on acid-free paper

This Springer imprint is published by Springer Nature
The registered company is Springer International Publishing AG
The registered company address is: Gewerbestrasse 11, 6330 Cham, Switzerland

Preface

The combination of different intelligent methods is a very active research area in artificial intelligence (AI). The aim is to create integrated or hybrid methods that benefit from each of their components. It is generally believed that complex problems can be easier solved with such integrated or hybrid methods.

Some of the existing efforts combine what are called soft computing methods (fuzzy logic, neural networks, and genetic algorithms) either among themselves or with more traditional AI methods such as logic and rules. Another stream of efforts integrates case-based reasoning or machine learning with soft computing or traditional AI methods. Yet another integrates agent-based approaches with logic and also non-symbolic approaches. Some of the combinations have been quite important and more extensively used, such as neuro-symbolic methods, neuro-fuzzy methods, and methods combining rule-based and case-based reasoning. However, there are other combinations that are still under investigation, such as those related to the Semantic Web. In some cases, combinations are based on first principles, whereas in other cases, they are created in the context of specific applications.

Important topics of the above area are (but not limited to) as follows:

- Case-Based Reasoning Integrations
- Genetic Algorithms Integrations
- Combinations for the Semantic Web
- Combinations and Web Intelligence
- Combinations and Web Mining
- Fuzzy-Evolutionary Systems
- Hybrid Deterministic and Stochastic Optimization Methods
- Hybrid Knowledge Representation Approaches/Systems
- Hybrid and Distributed Ontologies
- Information Fusion Techniques for Hybrid Intelligent Systems
- Integrations of Neural Networks
- Intelligent Agents Integrations
- Machine Learning Combinations
- Neuro-Fuzzy Approaches/Systems

- Applications of Combinations of Intelligent Methods to the following:
  - Biology and Bioinformatics
  - Education and Distance Learning
  - Medicine and Health Care

This volume includes extended and revised versions of some of the papers presented in the 5th International Workshop on Combinations of Intelligent Methods and Applications (CIMA 2015) and also papers submitted especially for this volume after a CFP. CIMA 2015 was held in conjunction with the 27th IEEE International Conference on Tools with Artificial Intelligence (ICTAI 2015).

We would like to express our appreciation to all authors of submitted papers as well as to the members of CIMA 2015 program committee for their excellent review work.

We hope that these post-proceedings will be useful to both researchers and developers.

Patras, Greece — Ioannis Hatzilygeroudis
Coventry, UK — Vasile Palade
Alexandroupoli, Greece — Jim Prentzas

Reviewers (CIMA 2015 Program Committee)

Ajith Abraham, Machine Intelligence Research Labs (MIR Labs)
Plamen Angelov, Lancaster University, UK
Nick Bassiliades, Aristotle University of Thessaloniki, Greece
Maumita Bhattacharya, Charles Sturt University, Australia
Kit Yan Chan, Curtin University, Australia
Gloria Cerasela Crisan, University "Vasile Alecsandri" of Bacau, Romania
Wei Fang, Jiangnan University, China
Foteini Grivokostopoulou, University of Patras, Greece
Ioannis Hatzilygeroudis, University of Patras, Greece (co-chair)
Constantinos Koutsojannis, T.E.I of Patras, Greece
Rudolf Kruse, University of Magdeburg, Germany
George Magoulas, Birkbeck College, UK
Ashish Mani, Dayalbagh Educational Institute, India
Antonio Moreno, University Rovira i Virgili, Spain
Vasile Palade, Coventry University, UK (co-chair)
Isidoros Perikos, University of Patras, Greece
Jim Prentzas, Democritus University of Thrace, Greece (co-chair)
David Sanchez, University Rovira i Virgili, Spain
Kyriakos Sgarbas, University of Patras, Greece
George A. Tsihrintzis, University of Piraeus, Greece
Douglas Vieira, ENACOM—Handcrafted technologies, Brazil

Contents

Chapter 1
Real-Time Investors' Sentiment Analysis from Newspaper Articles

Konstantinos Arvanitis and Nick Bassiliades

Abstract Recently, investor sentiment measures have become one of the more widely examined areas in behavioral finance. They are capable of both explaining and forecasting stock returns. The purpose of this paper is to present a method, based on a combination of a Naïve Bayes classifier and the n-gram probabilistic language model, which can create a sentiment index for specific stocks and indices of the New York Stock Exchange. An economic useful proxy for investor sentiment is constructed from U.S. news articles mainly provided by The New York Times. Initially, a large amount of articles for ten big companies and indices is collected and processed, in order to be able to extract a sentiment score from each one of them. Then, the classifier is trained from the positive, negative and neutral articles, so that it is possible afterwards to examine the sentiment of any unseen newspaper article, for any company or index. Subsequently, the classification task is tested and validated for its accuracy and efficiency. The widely used Baker and Wurgler sentiment index [2] is used as a comparison measure for predicting stock returns. In a sample of S&P 500 index from 2004 to 2010 on monthly basis, it is shown that the new sentiment index created has, on average, twice the predictive ability of Baker and Wurgler's index, for the existing time frame.

Keywords Sentiment analysis · Data mining · Sentiment index · Investor sentiment · Stock returns · Naïve bayes classifier · n-gram language model

K. Arvanitis · N. Bassiliades (✉)
Department of Informatics, Aristotle University of Thessaloniki, Thessaloniki, Greece
e-mail: nbassili@csd.auth.gr

K. Arvanitis
e-mail: konstantinos_ar@hotmail.com

I. Hatzilygeroudis et al. (eds.), *Advances in Combining Intelligent Methods*,
Intelligent Systems Reference Library 116, DOI 10.1007/978-3-319-46200-4_1

1.1 Introduction

News media is a very competitive industry whose main goal is to capture attention. Shiller [21] notes that news play a crucial role in buying or selling decisions among traders, who constantly react to new incoming information. He further argues that the news media are important players in creating market sentiment and similar thinking as it spreads ideas and, thus, can significantly contribute to herding behavior and influence price movement on financial markets. Behavioral finance supplements standard finance by introducing the revolutionary belief that behavior is not 'rational' but 'normal' [24]. If financial markets are not always rational then perhaps investors should take into account the psychology of the market. How this should be achieved has received great attention in the academic literature during the last decade. Most research tries to construct an index of investor sentiment with the help of various indicators. Baker and Wurgler [2] construct an index of investor sentiment that is based on the common variation in six proxies for sentiment: the closed-end fund discount, share turnover, the number and average first-day returns on IPO's, the equity share in new issues and the dividend premium.

As for ways to measure investor sentiment: there are direct and indirect measures. Direct measures are based on surveys taken from certain groups of people, for instance: global fund managers. Indirect measures are based on market data such as price and volatility. Both have their own merits and drawbacks. Investor sentiment is a much debated topic but it is not yet clear how it should be measured. Current literature attempts to capture investor sentiment by combining multiple imperfect proxies. Such a proxy is for example the market volatility index ('VIX'), which measures the implied volatility of options on the Standard and Poor's 100 stock index and is known as the 'investor fear gauge'. Popular directed measures of sentiment are different sort of confidence indices. In numerous countries and markets there are multiple indices available that try to track consumer or (retail) investor confidence by means of surveys.

Recently, research is focusing more on methods that capture sentiment with the help of media and computational linguistics. A freely available tool, called Google Search Volumes, is used by [12] to predict stock returns. Changes in volumes of words like "market crash" and "bear market" can predict stock returns while changes in positive search word volumes such as of "bull market" do not. One of the most easy and effective but less sophisticated ways for analyzing text is by means of a Bayesian approach. All that is required are two text files that represent negative and positive words or sentences. Then, a specific text can be classified as negative or positive depending on the similarity with the two basis files. This method was tested by [14], which successfully determined if a movie was regarded as good or bad. A major drawback of this method is the fact it only classifies text into positive/negative, without being able to determine the degree of positivity/negativity. Another drawback is that the quality of the basis files determines the quality of the analysis.

In this paper a useful proxy for investor sentiment is constructed with the help of financial news from U.S. newspapers from 2004 to 2014. The construction of the sentiment index follows a variation of a Bayesian approach, combining a Naïve Bayes classifier with the n-gram probabilistic language model that is based on Markov chains. The classifier is trained from three highly targeted text lists containing positive, negative and, also, neutral text acquired from the newspaper articles. The main objective is that a sentiment index could be constructed for any ticker of the U.S. Stock Exchange in real-time, in order to help investors classify stocks or measure the overall market sentiment.

In contrast to existing literature, our analysis is much broader given that a sentiment index can be created for any company or index of the stock exchange. The results in many of the previous studies where Twitter is used as data source suffer from noise, since many Tweets are insignificant but affect the overall result. Our sentiment index is created from official news feed and articles of the New York Times and the result is much more factual and clear. In other studies, there is the limit of the research area which only deals with some indices, while others have the limit of the time frame, which has to be many days or months in contrast with our approach where you can create the sentiment index on a daily, weekly, monthly or annual time frame. In addition, we manage to tackle the issue of the absence of the degree of positivity and negativity that other studies have, by adding a list of neutral articles, twice the size of the positive and negative articles. This, combined with a safe threshold, allows us to classify correctly new articles with an actual sentiment impact. Furthermore, the quality of the basis files does not affect the quality of the analysis, in our approach, since via the self-learning feature of the classifier, our method adds previously classified text fragments to the basis files.

The paper is structured as follows: Sect. 1.2 provides the appropriate background knowledge to our work, Sect. 1.3 reports some related work, Sect. 1.4 presents the data sources and the methodology, Sect. 1.5 continues with the empirical findings and finally, Sect. 1.6 concludes.

1.2 Background

1.2.1 Framing Effects

The framing effect is an example of cognitive bias, in which people react to a particular choice in different ways depending on how it is presented; e.g. as a loss or as a gain [15]. People tend to avoid risk when a positive frame is presented but seek risks when a negative frame is presented [26].

Framing effects within the news media have been an important research topic among journalism, political science and mass communication scholars. Price et al. argue [17] that the news framing effect has to do with the way events and issues are packaged and presented by journalists to the public. They believe that news frames can fundamentally affect the way readers understand events and issues. Authors suggest that news frames can activate certain ideas, feelings, and values, encourage particular trains of thoughts and lead audience members to arrive at predictable conclusions.

Price and Tewksbury [16] explain the news media framing effect by using the applicability effect in their knowledge activation process model. A framing effect of a news story renders particular thoughts applicable through salient attributes of a message such as its organization, selection of content or thematic structure. The knowledge activation model assumes that at any particular point in time, a mix of particular items of knowledge that are subject to processing (activation) depends on characteristics of a person's established knowledge store. When evaluating situations, people tend to use (activate) ideas and feelings that are most accessible and applicable.

Iyengar [11] examines the impact of news framing on the way people ascribe responsibility for social, political, and economic conditions. He finds that media more often take an episodic rather than a thematic perspective towards the events they cover.

Vliegenthart et al. [27] investigate the effect of two identified news frames, risk and opportunity, on public support regarding the enlargement of the European Union. They find that participants in the opportunity frame condition show significantly higher support compared to participants in the risk condition.

These studies show that framing influences the perception of new information and may be a powerful tool in influencing public opinion and, as a consequence, the public's future actions. Casual observation suggests that the content of news about the stock market could be linked to investor psychology and sociology. However, it is unclear whether the financial news media induces, amplifies, or simply reflects investors' interpretations of stock market performance.

1.2.2 Investor Sentiment Proxy Construction

Investor sentiment is a much debated topic but it is not yet clear how it should be measured. Current literature attempts to capture investor sentiment by combining multiple imperfect proxies. Such a proxy is for example the market volatility index ('VIX'), which measures the implied volatility of options on the Standard and Poor's 100 stock index and is known as the 'investor fear gauge'. The VIX index is often used as a contrarian indicator in that extreme levels indicate market turning points and is supported by the theory of market over- and under reaction. Popular directed measures of sentiment are different sort of confidence indices. In numerous countries and markets there are multiple indices available that try to track consumer or (retail) investor confidence by means of surveys.

Two widely known indices for U.S. consumer confidence are the Conference Board's Consumer Confidence Index (CCI) and the University of Michigan's Index of Consumer Sentiment. Bram and Ludvigson [5] found the former is better at explaining most categories of consumer spending. Qiu and Welch [18] find that consumer confidence can be a good proxy for investor sentiment and plays a robust role in financial market pricing.

Tumarkin and Whitelaw [25] use the opinions and views of message board users for examining the relationship between sentiment and abnormal stock returns and trading volume. Although investor opinion correlates with abnormal industry-adjusted returns they find no evidence contrary to market efficiency.

Bollen et al. [4] analyze large amounts of tweets (short bursts of inconsequential information) for mood swings. They use two tools that determine mood with the help of computational linguistics: OpinionFinder and Google-Profile of Mood States (GPMOS). OpinionFinder determines positive versus negative moods and GPMOS measures mood in six dimensions (Calm, Kind, Happy, Vital, Alert and Sure). They claim that the daily closing price of the DJIA can be predicted 4 days ahead with 87.6 % accuracy. However, this cannot be verified because the GPMOS is not made public. OpinionFinder is an open source project of several American Universities and identifies positive/negative words, actions and subjective/objective statements. Its developers claim to accurately classify polarity about 74 % of the time.

While OpinionFinder is a clear improvement over a Bayesian method it still lacks the ability to determine the degree of negativity/positivity. A company called OpenAmplify3 claims to have successfully resolved this problem. Although their method is black box we can analyze the input and output of their service. Open-Amplify requires English text files as input and returns output with the help of an application programming interface (API). Their analysis is quite extensive and can be divided into five main categories: topics, actions, styles, demographics and topic intentions analysis. Topic analysis is done on a co-reference basis, meaning that different words can be identified as belonging to the same topic. For instance: 'Jack' and 'Jill' and 'He' and 'She' are connected but also 'Coca Cola Company' and 'CCC' are linked together. Every topic scores a degree of polarity (negativity/ positivity) on a scale of –1 to 1 where the former indicates extreme negativity and the latter indicates extreme positivity. Overall text polarity is the weighted average polarity of all separate topics. Weighting is done based on a relevance score. Topics that are weakly related to all other topics are given a low relevance score and have low impact on overall text polarity. This is possible because OpenAmplify identifies relationships between topics and organizes them into a broad range of domains. Other interesting features are action analysis, contrast (degree of certainty) and temporality (timeframe). Given that OpenAmplify works, having an extensive analysis of a large amount of news articles gives you the possibility to construct a wide variety of (investor) sentiment proxies. For example: irrelevant text can be filtered out by focusing on the domain 'business' with subdomain 'stock market'. The average polarity of the remaining text can be a proxy for investor sentiment.

1.3 Related Work

Previous research investigates the immediate impact news media might have on the performance of financial markets. For instance, Antweiler and Frank [1] investigate the effect of Internet stock message boards posted on the websites of Yahoo! Finance and Raging Bull on the short-term market performance of 45 U.S. listed companies. They find weak evidence that the number of content messages posted helps to predict stock's intraday volatility but do not find evidence of news media content in-between the content of the Wall Street Journal column Abreast of the Market and the stock market on a daily basis. They, also, find that unusually low or high values of media pessimism predict high trading volume, while low market returns lead to high media pessimism, and conclude that news media content can serve as a proxy for investor sentiment. In a more recent study, Garcia [10] constructs a daily proxy for investor sentiment by taking a fraction of negative and positive words in two columns of financial news, Financial Markets and Topics in Wall Street from the New York Times. He finds evidence of an asymmetric predictive activity of news content on stock returns, especially during recessions. The effect is particularly strong on Mondays and on trading days after holidays, which persists into the afternoon of the trading day.

While some trading in the market brings noise traders with different models who cancel each other out, a substantial percentage of trading strategies are correlated, leading to aggregate demand shifts. As Shleifer and Summers elaborate [22], the reason for this is that the judgmental biases affecting investors in information processing tend to be the same. For example, subjects in psychological experiments tend to make the same mistake; they do not make random mistakes. Indeed, Barber et al. [3] utilize brokerage data and find that individual investors predominantly buy the same stocks as each other contemporaneously, and that this buying pressure drives prices upwards. Similarly, Schmeling [19] employs survey data and finds that individual investor sentiment forecasts stock market returns. In effect, these studies reveal that arbitrageurs are not always successful in bringing prices back in line with fundamentals. Thus, shifts in the demand for stocks that are independent of fundamentals may persist, and thus be observable.

Dickinson and Hu [8] seek to predict a sentiment value for stock related tweets on Twitter, and demonstrate a correlation between this sentiment and the movement of a company's stock price in a real time streaming environment. They use both *n*-gram and "word2vec"[1] textual representation techniques alongside a random forest classification algorithm to predict the sentiment of tweets. These values are then evaluated for correlation between stock prices and Twitter sentiment for that each company. The results show significant correlations between price and sentiment for several individual companies. Some companies such as Microsoft and Walmart show strong positive correlation, while others such as Goldman Sachs and

[1]https://code.google.com/p/word2vec .

Cisco Systems show strong negative correlation. This suggests that consumer facing companies are affected differently than other companies.

Das and Chen [6] developed a methodology for extracting small investor sentiment from stock message boards. Their findings showed that five distinct classifier algorithms coupled by a voting scheme are found to perform well against human and statistical benchmarks. Also, they state that time series and cross-sectional aggregation of message information improves the quality of the sentiment index. Their empirical applications evidence a relationship with stock returns, on a visual level, by phase-lag analysis, using pattern recognition and regression methods. Last but not least, they state that sentiment has an idiosyncratic component, and aggregation of sentiment across stocks tracks index returns more strongly than with individual stocks.

Sehgal and Song [20] introduce a novel method to predict sentiment about stock using financial message boards. They state that web financial information is not always reliable and for this reason they propose a new measurement known as TrustValue which takes into account the trustworthiness of an author. In their work, it is shown that TrustValue improves prediction accuracy by filtering irrelevant or noisy sentiments. Sentiment and TrustValue are used together to make the model for stock prediction. They used the intuition that sentiments effect stock performance over short time period and they captured this with Markov model. Their stock prediction results showed that sentiment and stock value are closely related and web sentiment can be used to predict stock behavior with seasonable accuracy.

The linear causality framework is widely adopted in the behavioral finance literature when evaluating the predictive content that sentiment may have upon stock returns. Dergiades [7] finds out that there is reasonable statistical evidence to support that sentiment embodies significant predictive power with respect to stock returns. His study contributes to the understanding of the non-linear causal linkage between investors' sentiment dynamics and stock returns for the US economy, by employing the sentiment index developed by Baker and Wurgler and within a non-linear causality framework.

1.4 News Articles Classification Methodology and Sources

In this section, we present the sources that were used for this work; the methodology we followed and the processing the data went through. The key concept in this work is to train a classifier which is the most appropriate to classify articles with financial content about companies (as positive, negative or neutral).

Figure 1.1 shows diagrammatically the processes and the methods used for the data extraction, storage and the preprocessing in order to construct the three lists from which the classifier is trained using *n*-gram language models. As previously mentioned, except for positive and negative categories, a text can also be classified as neutral so that the result would be more accurate with reduced noise.

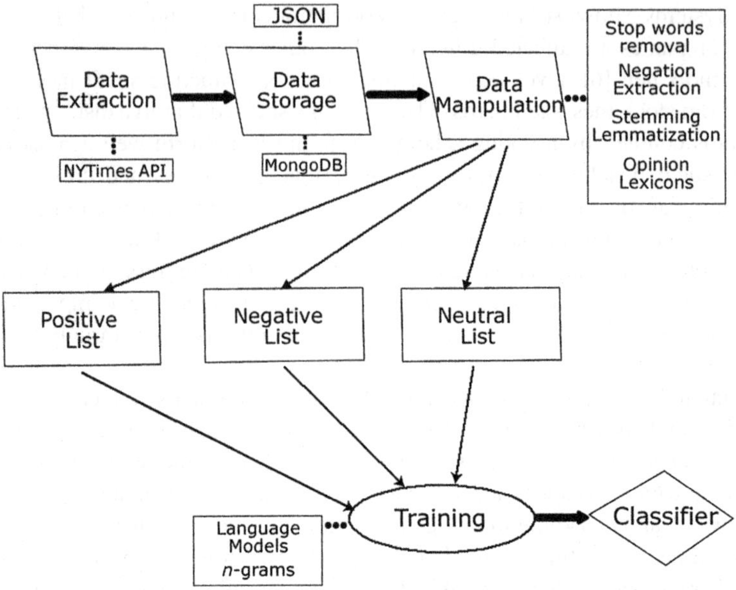

Fig. 1.1 Architecture and workflow of our methodology

1.4.1 News Sources and Preprocessing

Newspaper articles used for the analysis were obtained from the New York Times with NYT Article Search API v2,[2] which can search articles from 1851 to today, retrieving headlines, abstracts, lead paragraphs and other article metadata.

Search requests follow a standard URI structure whose main parameter is the search query term which is being searched on the article body, headline and byline. The structure of a search request is the following:

http://api.nytimes.com/svc/search/v2/articlesearch.response-format?[q=search term&fq=filter-field:(filter-term)&additional-params=values]&api-key=####

Besides the search query term, a filtered search query feature is applied, which uses standard Lucene[3] syntax and can specify the fields and the values that a query will be filtered on. Lucene syntax separates the filter field name and value with a colon, and surrounds multiple values with parentheses, like the following:

```
&fq = news_desk: ("Sports" "Foreign") AND glocations:
("NEW YORK CITY")
```

[2]http://developer.nytimes.com/docs/read/article_search_api_v2.

[3]https://lucene.apache.org/.

In this way, the scope of the search can be narrowed and the articles returned would be more accurate, which in this work involve exclusively business or financial topics.

All articles are returned in JSON[4] format and stored in MongoDB[5] database. News data for each company is stored in a different collection so that all bulk of data is clustered and easy to manipulate. All collections are sorted by ascending order according to the publish date of the articles. New articles for a company are stored in the corresponding collection in chronological order. The initial look of an article is shown in Fig. 1.2, which was obtained from New York Times website. The JSON format of the article in Fig. 1.2 has the structure shown in Fig. 1.3.

For the opinion lexicons, two positive and negative dictionary files are used [13], which are useful for textual analysis in financial applications. Phrases like 'not good' are converted to '!good' and added at the corresponding dictionary to distinguish negation.

Another widely used feature of natural language processing is used, which has to do with removing the stop words from the examined text, so that the text classification can be applied to only the words that really count and have positive or negative effect to the overall sentiment. The Stop Word Lists used in the analysis [13] are divided in five categories: Generic, Names, Geographic, Currencies, Dates and Numbers. Besides the removal of stop words, the training procedure consists of another feature which is to collapse all the different inflectional forms of a lemma to its base dictionary form, which can be found in the lexicons. Classification was tested in different ways towards the stop word lists, like excluding some of them or applying the classification to the text without removing any stop words. The tests showed that the best performance was achieved by removing all the stop words from the text and leaving only words that have sentiment impact.

Since the text gets a form, which is optimal and easy to extract a safe score, it is passed to the lexicons to count the occurrence of each word in the text that exist in any of the lexicons. If a word in the text belongs to the positive lexicon, the counter of the sentiment score is increased by one and if it belongs to the negative lexicon, the counter is decreased by one. Finally, a sentiment score about the examined text is obtained, which must be used to categorize it as positive, negative or neutral. For accuracy reasons, a threshold is set for sentiment score of higher than 2, for positive, and lower than -2, for negative, and all between them are categorized as neutral.

Three lists are created, one for positive, one for negative and one for neutral articles. For our study, we have 5000 positive articles, 5000 negative articles and 10000 neutral articles. These lists are used to train the classifier, which is a Dynamic Language Model classifier that uses n-gram language models, as explained in the next section. Training is based on a multivariate estimator for the category distribution and dynamic language models for the per-category character sequence estimators. It calculates conditional and joint probabilities of each category for the

[4]http://json.org/ .

[5]https://www.mongodb.org/.

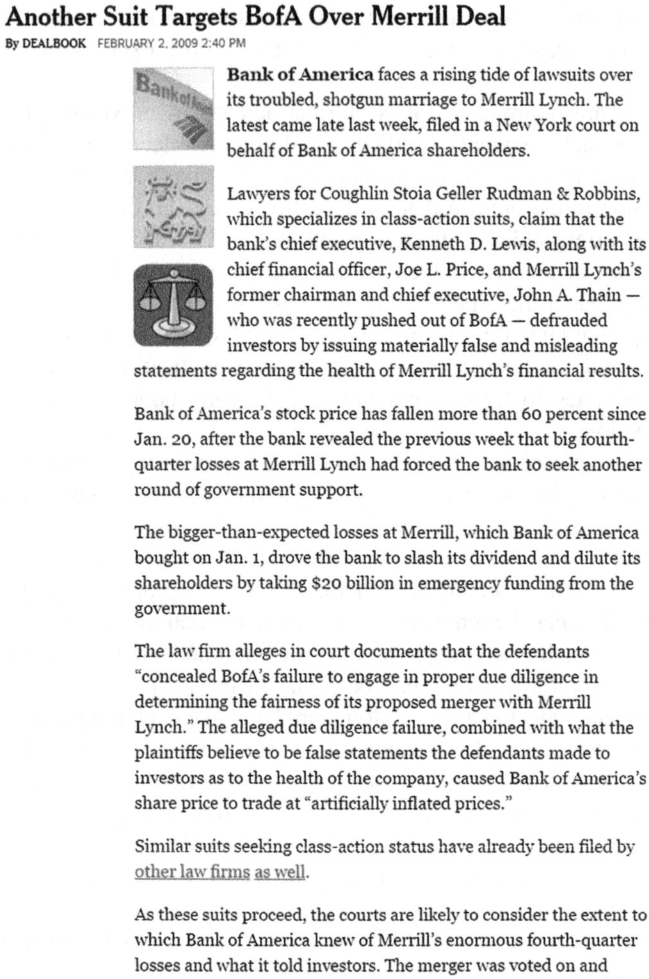

Another Suit Targets BofA Over Merrill Deal

By DEALBOOK FEBRUARY 2, 2009 2:40 PM

Bank of America faces a rising tide of lawsuits over its troubled, shotgun marriage to Merrill Lynch. The latest came late last week, filed in a New York court on behalf of Bank of America shareholders.

Lawyers for Coughlin Stoia Geller Rudman & Robbins, which specializes in class-action suits, claim that the bank's chief executive, Kenneth D. Lewis, along with its chief financial officer, Joe L. Price, and Merrill Lynch's former chairman and chief executive, John A. Thain — who was recently pushed out of BofA — defrauded investors by issuing materially false and misleading statements regarding the health of Merrill Lynch's financial results.

Bank of America's stock price has fallen more than 60 percent since Jan. 20, after the bank revealed the previous week that big fourth-quarter losses at Merrill Lynch had forced the bank to seek another round of government support.

The bigger-than-expected losses at Merrill, which Bank of America bought on Jan. 1, drove the bank to slash its dividend and dilute its shareholders by taking $20 billion in emergency funding from the government.

The law firm alleges in court documents that the defendants "concealed BofA's failure to engage in proper due diligence in determining the fairness of its proposed merger with Merrill Lynch." The alleged due diligence failure, combined with what the plaintiffs believe to be false statements the defendants made to investors as to the health of the company, caused Bank of America's share price to trade at "artificially inflated prices."

Similar suits seeking class-action status have already been filed by other law firms as well.

As these suits proceed, the courts are likely to consider the extent to which Bank of America knew of Merrill's enormous fourth-quarter losses and what it told investors. The merger was voted on and

Fig. 1.2 Example of a New York Times article

classified object and the classifier returns one best category as result of classification process. Experimental results show that using language models in classification, we are able to obtain better performance than traditional Naïve Bayes classifier.

1.4.2 Classification Methodology

At this point, we describe the methods adopted for the classification of the news articles and the validation check of the methodology. The classification procedure uses *n*-gram language models and it is considered as an extension of the traditional Naïve Bayes classifier, with the difference that the Laplace smoothing is replaced

```
{
 "response": {
   "meta": {
     "hits": 20,
     "time": 385,
     "offset": 0   },
   "docs": [
     {
       "web_url": "http:\/\/dealbook.nytimes.com...",
       "snippet": "Bank of America faces ...",
       "lead_paragraph": "Bank of America faces ...",
       "abstract": "Bank of America faces [...]",
       "print_page": null,
       "blog": [       ],
       "source": "The New York Times",
       "multimedia": [       ],
       "headline": {
         "main": "Another Suit Targets ...",
         "kicker": "DealBook"       },
       "keywords": [
         {
           "rank": "1",
           "name": "type_of_material",
           "value": "News"          }
       ],
       "pub_date": "2009-02-02T14:40:29Z",
       "document_type": "blogpost",
       "news_desk": null,
       "section_name": "Business Day",
       "subsection_name": null,
       "byline": {
         "person": [
           {
             "organization": "",
             "role": "reported",
             "rank": 1          }
         ],
         "original": "By DEALBOOK"
       },
       "type_of_material": "Blog",
       "_id": "4fd394388eb7c8105d8c8fdd",
       "word_count": 512     }
   ]
 },
 "status": "OK",
 "copyright": "Copyright (c) 2013..."
}
```


Fig. 1.3 JSON format of a New York Times article

by some more sophisticated smoothing methods. A Naïve Bayes classifier assumes that the value of a particular feature is unrelated to the presence or absence of any other feature, given the class variable. For instance, a vehicle may be considered to be a bike if it has two wheels, no doors, and has about 150 cm wheelbase. A Naïve

Bayes classifier considers each of these features to contribute independently to the probability that this vehicle is a bike, regardless of the presence or absence of the other features. Experimental results show that using a language model, we are able to obtain better performance than traditional Naïve Bayes classifier. Language models have been successfully applied in many application areas such as speech recognition and statistical natural language processing.

In recent years, it is confirmed that they are also an attractive approach for Information Retrieval (IR) such as the query likelihood model, because they can provide effectiveness comparable to the best state of the art systems. As a result of this fact, language models are used to other IR-related tasks, such as tracking, topic detection and classification. In this work, an attempt is being made to use language models in text classification, specifically from newspaper articles with financial content about companies, evaluate the accuracy of the method and compare the new sentiment index created with another widely used index.

Language modeling aims to predict the probability of natural word sequences. More simply, the goal is to put high probability on word sequences that actually occur and low probability on the ones that never occur. Given a word sequence $w_1 w_2 \ldots w_T$ to be used as a test corpus, the quality of a language model can be measured by the empirical perplexity (or entropy) on this corpus:

$$Perplexity = \sqrt[T]{\frac{1}{P(w_1 \ldots w_T)}} \tag{1.1}$$

$$Entropy = \log_2(Perplexity) \tag{1.2}$$

The main objective is to obtain a small perplexity. The simplest and most successful basis for language modeling is the n-gram model: Note that by the chain rule of probability we can write the probability of any sequence as

$$P(w_1 w_2 \ldots w_T) = \prod_{i=1}^{T} P(w_i | w_1 \ldots w_{i-1})$$

An n-gram model approximates this probability by assuming that the only words relevant to predicting $P(w_i | w_1 \ldots w_{i-1})$ are the previous $n - 1$ words; that is, it assumes the Markov n-gram independence assumption

$$P(w_i | w_1 \ldots w_{i-1}) = P(w_{i-n+1} | w_1 \ldots w_{i-1})$$

A straightforward maximum likelihood estimate of n-gram probabilities from a corpus is given by the observed frequency

$$P(w_{i-n+1} | w_1 \ldots w_{i-1}) = \frac{\#(w_{i-n+1} \ldots w_i)}{\#(w_{i-n+1} \ldots w_{i-1})}$$

where #(.) is the number of occurrences of a specified gram in the training corpus. Unfortunately, using grams of length up to n entails estimating the probability of W^n events, where W is the size of the word vocabulary. This fact makes it necessary to choose a relatively smaller n (beyond 2 to 7). In addition, it is likely to encounter novel n-grams that were never witnessed during training, because of the heavy tailed nature of language (i.e. Zipf's law). Therefore, a mechanism for assigning non-zero probability to novel n-grams is needed. One standard approach to cope with potentially missing n-grams is to use some sort of back-off estimator, which is relatively simple and has the following form:

$$P(w_i|w_{i-n+1}\ldots w_{i-1}) = \begin{cases} \hat{P}(w_i|w_{i-n+1}\ldots w_{i-1}), & if\#(w_{i-n+1}\ldots w_i) > 0 \\ \beta(w_{i-n+1}\ldots w_{i-1}) \times P(w_i|w_{i-n+2}\ldots w_{i-1}), & otherwise \end{cases}$$

where

$$\hat{P}(w_i|w_{i-n+1}\ldots w_{i-1}) = \frac{discount\#(w_{i-n+1}\ldots w_i)}{\#(w_{i-n+1}\ldots w_{i-1})} \tag{1.3}$$

is the discounted probability, and $\beta(w_{i-n+1}\ldots w_{i-1})$ is a normalization constant calculated to be

$$\beta(w_{i-n+1}\ldots w_{i-1}) = \frac{1 - \sum_{x:\#(w_{i-n+1}\ldots w_{i-1}x)>0} \hat{P}(x|w_{i-n+1}\ldots w_{i-1})}{1 - \sum_{x:\#(w_{i-n+1}\ldots w_{i-1}x)>0} \hat{P}(x|w_{i-n+2}\ldots w_{i-1})}$$

An n-gram is first matched against the language model to see if it has been observed in the training corpus. If that fails, the n-gram is then reduced to an $n-1$-gram by shortening the context by one word. The discounted probability (Eq. 1.3) can then be computed using different smoothing approaches. Smoothing techniques are analyzed further below (Eq. 1.5).

Text classifiers, like dynamic language model classifiers, attempt to identify attributes, which distinguish documents in different categories. Vocabulary terms, local n-grams, word average length, or global syntactic and semantic properties may be such attributes. Also, Language models provide another natural avenue to constructing text classifiers as they attempt to capture such regularities. An n-gram language model can be applied to text classification in a similar manner to a Naïve Bayes model. That is, we categorize a document according to

$$c^* = argmax\{P(c|d)\}$$

Using Bayes rule, this can be rewritten as

$$\begin{aligned} c^* &= argmax\{P(c)P(d|c)\} \\ &= argmax\left\{P(c)\prod_{i=1}^{T} P(w_i|w_{i-n+1}\ldots w_{i-1},c)\right\} \\ &= argmax\left\{P(c)\prod_{i=1}^{T} P_c(w_i|w_{i-n+1}\ldots w_{i-1})\right\} \end{aligned} \tag{1.4}$$

Here, $P(d|c)$ is the likelihood of d under category c, which can be computed by an n-gram language model. Likelihood is related to perplexity and entropy by Eqs. (1.1) and (1.2). $P_c(w_i|w_{i-n+1}\cdots w_{i-1})$ is computed using back-off language models which are learned separately for each category by training on a data set from that category. Then, to categorize a new document d, the document is supplied to each language model, the likelihood (or entropy) of d under the model is evaluated, and the winning category is picked according to Eq. (1.4).

The n-gram, which is a subsequence of length n of the items given, has a certain size that needs to be set for the Language Model classifier algorithm. The Language Model rule is to classify a newly given document based on prediction occurring n-grams. The algorithm uses a word based n-gram to classify articles so an appropriate size should be the average length of a sentence.

If we take into account that the traditional Naïve Bayes classifier is a unigram classifier with Laplace smoothing, then it is obvious that n-gram classifiers are in fact a straightforward generalization of Naïve Bayes. However, n-gram language models possess many advantages over Naïve Bayes classifiers, for larger n, including modelling longer context and exploiting better smoothing techniques in the presence of sparse data. Another notable advantage of the language modelling based approach is that it does not incorporate an explicit feature selection procedure. For Naïve Bayes text classifiers, features are the words, which are considered independent of each other given the category. Instead, Language Model classifiers consider all possible n-grams as features. Their importance is implicitly considered by their contribution to the quality of language modelling. The over-fitting problems associated with the subsequent feature explosion are nicely handled by applying smoothing techniques like Laplace smoothing.

Two general formulations are used in smoothing: back-off and interpolation. Both smoothing methods can be expressed in the following general form:

$$P(w|c_i) = \begin{matrix} P_s(w|c_i) & w\, is\, seen\, in\, c_i \\ a_{c_i}P_u(w|C)' & w\, is\, unseen\, in\, c_i \end{matrix} \tag{1.5}$$

This form shows that for a class c_i, one estimate is made for the words seen in the class, and another estimate is made for the unseen words. In the second case, the estimate for unseen words is based on the entire collection, i.e., the collection model. The zero-probability problem is solved by incorporating the collection model, which also generates the same effect as the IDF factor [23] that is commonly used in IR [9].

The accuracy of the classification is estimated by applying a popular method in machine learning, called *k*-fold cross-validation. Estimating the accuracy of a classifier induced by supervised learning algorithms is important not only to predict its future prediction accuracy, but also for choosing a classifier from a given set (model selection), or combining classifiers [28]. An estimation method with low bias and low variance is the best fit to estimate the final accuracy of a classifier.

A classifier is a function that maps an unlabeled instance to a label using internal data structures. An inducer, or an induction algorithm, builds a classifier from a given dataset. Let V be the space of unlabeled instances and Y the set of possible labels. Let $X = V \times Y$ be the space of labeled instances and $D = \{x_1, x_2, \ldots, x_n\}$ be a dataset (possibly a multiset) consisting of n labeled instances, where $x_i = \langle u_i \in V, y_i \in Y \rangle$. A classifier C maps an unlabeled instance $v \in V$ to a label $y \in Y$ and an inducer I maps a given dataset D into a classifier C. The notation $I(D, v)$ will denote the label assigned to an unlabeled instance v by the classifier built by inducer I on dataset D, *i.e.*, $I(D, v) = (I(D))(v)$.

The accuracy of a classifier C is the probability of correctly classifying a randomly selected instance, *i.e.*, $acc = Pr(C(v) = y)$ for a randomly selected instance $\langle u, y \rangle \in X$, where the probability distribution over the instance space is the same as the distribution that was used to select instances for the inducer's training set. Given a nite dataset, the future performance of a classifier induced must be estimated by the given inducer and dataset. A single accuracy estimate is usually meaningless without a confidence interval, so such an interval should be approximated when possible. Also, in order to identify weaknesses the cases where the estimates fail should be identified.

In *k*-fold cross-validation, sometimes called rotation estimation, the dataset D is randomly split into k mutually exclusive subsets (the folds) $D_1, D_2, \ldots, D_k$ of approximately equal size. The inducer is trained and tested k times; each time $t \in \{1, 2, \ldots, k\}$, it is trained on D/D_t and tested on D_t. The cross-validation estimate of accuracy is the overall number of correct classifications, divided by the number of instances in the dataset. Formally, let $D_{(i)}$ be the test set that includes instance $x_i = \langle v_i, y_i \rangle$, then the cross-validation estimate of accuracy

$$acc_{cv} = \frac{1}{n} \sum_{\{u_i, y_i\} \in D} \delta(I(D/D_{(i)}, u_i), y_i) \tag{1.6}$$

The cross-validation estimate is a random number that depends on the division into folds. In cross-validation, it is useful to obtain an estimate for many performance indicators such as accuracy, precision, recall, or F-score. In most cases, the accuracy of a classifier is estimated in a supervised-learning environment. In such a setting, there is a certain amount of labeled data and the goal is to predict how well a certain classifier would perform if this data is used to train the classifier and subsequently ask it to label unseen data. In 10-fold cross-validation, the 90 % of the data is repeatedly used to build a model and the remaining 10 % to test its accuracy. The average accuracy of the repeats is an underestimate for the true accuracy. Generally, this estimate is reliable, especially if the amount of labeled data is large enough and if the unseen data follows the same distribution as the labeled examples.

1.5 Results and Discussion

For classification tasks, the terms true positives, true negatives, false positives and false negatives (also Type I and Type II errors) compare the results of the classifier under test with trusted external judgments. The terms *positive* and *negative* refer to the classifier's prediction (sometimes known as the *expectation*), and the terms *true* and *false* refer to whether that prediction corresponds to the external judgment (sometimes known as the *observation*).

Accuracy is the overall correctness of the model and is calculated as the sum of correct classifications divided by the total number of classifications.

$$Accuracy = \frac{tp + tn}{tp + tn + fp + fn}$$

Precision is a measure of the accuracy, provided that a specific class has been predicted. It is defined by:

$$Precision = \frac{tp}{tp + fp}$$

where *tp* and *fp* are the numbers of true positive and false positive predictions for the considered class.

Recall is a measure of the ability of a prediction model to select instances of a certain class from a data set. It is commonly also called sensitivity, and corresponds to the true positive rate. It is defined by the formula:

$$Recall = Sensitivity = \frac{tp}{tp + fn}$$

where *tp* and *fn* are the numbers of true positive and false negative predictions for the considered class. Notice that $tp + fn$ is the total number of test examples of the considered class.

F-measure or balanced *F*-score is the harmonic mean of precision and recall.

$$F = 2 * \frac{precision * recall}{precision + recall}$$

Table 1.1 Performance indicators

	TP rate (recall)	*FP* rate	Accuracy	Precision	*F*-score
S&P 500	0.79	0.15	0.82	0.84	0.81
Dow Jones	0.88	0.06	0.92	0.90	0.89
Google	0.97	0.42	0.87	0.88	0.92
Bank of America	0.97	0.06	0.90	0.90	0.88
Apple	0.97	0.35	0.89	0.89	0.93
Ebay	0.96	0.31	0.90	0.91	0.93
Nike	0.97	0.49	0.90	0.92	0.94
Citigroup	0.84	0.03	0.91	0.94	0.89
Amazon	0.98	0.45	0.91	0.91	0.95
Microsoft	0.93	0.30	0.86	0.87	0.90
Average	0.93	0.26	0.89	0.90	0.91

While error rate or accuracy dominates much of the classification literature, *F*-measure is the most popular metric in the text classification and information retrieval communities. The reason is that typical text mining corpora have many classes and suffer from high class imbalance. Accuracy tends to undervalue how well classifiers are doing on smaller classes, whereas *F*-measure balances *precision* and *recall.*

After obtaining the 10 classifiers created by the 10-fold cross-validation on the training newspaper data, each one of them is evaluated at the corresponding test data set and the performance indicators are recorded which are shown in Table 1.1.

As Table 1.1 shows, the average scores that the classifier can achieve are 0.93 for recall, 0.89 for accuracy, 0.90 for precision and 0.91 for *F*-Score. Another noticeable thing is that the variance for those four measures is 0.004, 0.00082, 0.00072, 0.00148 respectively and the standard deviation is as low as 0.06 for recall, 0.029 for accuracy, 0.027 for precision and 0.039 for *F*-Score, which means that the classification procedure performs quite well in all cases, regardless of the index or the company. In addition, most of the scores are close except for those that correspond to the news articles for the main index S&P 500, which has approximately 4–5 times bigger data size than most of the other company tickers. For the Dow Jones Industrial Index, the classification performs quite well despite its big size.

In order to have a benchmark measure, investor sentiment data is used which was provided by Baker and Wurgler [2]. Baker and Wurgler created a sentiment index, which was updated in May 16, 2011, based on first principal component of six (standardized) sentiment proxies over 1962–2005 data, where each of the proxies has first been orthogonalized with respect to a set of macroeconomic conditions.

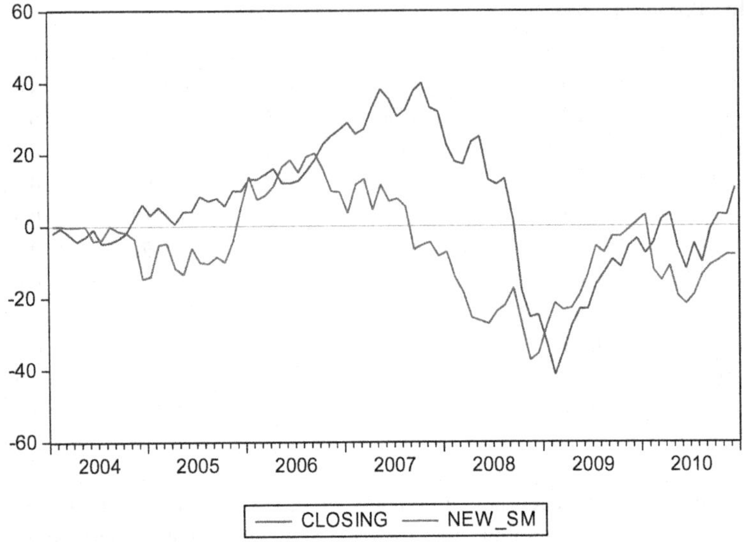

Fig. 1.4 Comparison chart of closing prices and new sentiment index

At first, chronological line-charts of sentiment analysis are created, for both annual and monthly time frames, for all the companies and indices that news data was collected. Afterwards, two charts of Baker and Wurgler are exported for annual and monthly time frames for the S&P 500 index from 2004 to 2010, and one more monthly chart with the closing prices of the index and all the charts are compared respectively. Furthermore, the new monthly sentiment line-chart of the S&P 500 index is compared with the one that contains the closing prices (Fig. 1.4).

Figure 1.4 reveals the possible co-movement of the new sentiment index and the closing prices in monthly basis, as it is obvious that the blue line containing the closing prices follows the curve of the red line, which is the sentiment index, sometime later than the first one.

The next step is to examine the new sentiment index on its ability to explain returns. This is done by applying a regression model. We examine the results provided by regressing the sentiment index on monthly data for the S&P 500 index for the period of January 2004 to December 2010.[6] In addition, we run the same specification using the sentiment index created by Baker and Wurgler for the same period. The first step of the regression model is to create an equation, which involves three variables: *Returns*, which is the dependent variable, and the *Sentiment index* and the *P/E* ratio, which are the regressors. *P/E* ratio is used here as a

[6]https://research.stlouisfed.org/fred2/series/SP500/downloaddata.

Table 1.2 Results of regression model for the new Sentiment index

Variable	Coefficient	Probability
New sentiment index	0.000896	0.0008
P/E ratio	0.000322	0.0683
C	–0.001405	0.8398
R-squared	0.152172	

Table 1.3 Results of regression model for Baker and Wurgler's Sentiment index

Variable	Coefficient	Probability
Baker and Wurgler index	–0.052797	0.0428
P/E ratio	9.09E-05	0.6540
C	–0.001421	0.8482
R-squared	0.074445	

variable related to the fundamentals of the index. The results in Table 1.2 show that the new sentiment index is very significant as its *p*-value is lower than 0.05 and very close to zero. *R*-squared is 0.152, which means that 15.2 % of variation in the dependent variable, which is the returns, can be explained by the new sentiment index and *P/E* ratio jointly.

On the other hand, the results in Table 1.3 show that Baker and Wurgler's sentiment index is marginally significant as it is close to 0.05. *R*-squared is 0.074, so only 7.4 % of variation in the returns can be explained by the sentiment index and *P/E* ratio jointly.

The final step is to apply simple rolling regression. The window size is set to 60, which is the months, in order to have a rolling 5-year time frame of the sentiment indices and the step size to 1 and we store the *P*-values and the *R*-squareds, so that we can then make the comparison graph with the ones from Baker and Wurgler's index.

Figure 1.5 shows the rolling p-values for the new sentiment index and Baker and Wurgler's sentiment index. As we can see the red *p*-values of the new index are almost every time close to zero, while the blue ones of Baker and Wurgler's index are a lot higher. This means that the new sentiment index is most of the time very significant for the equation and in any case, more significant than Baker and Wurgler's index.

In the next figure (Fig. 1.6) the rolling *R*-squareds for the new sentiment index and Baker and Wurgler's sentiment index are presented. In all cases the red one is above the blue which means it can predict better the future returns. The red has a peak at ~33 % while the blue at ~28 % and approximately the mean *R*-squared of the red is 20 % while the blue has 10 %, the half of the red. This fact shows in simple terms, that on average the new sentiment index has twice the predictive ability of Baker and Wurgler's index.

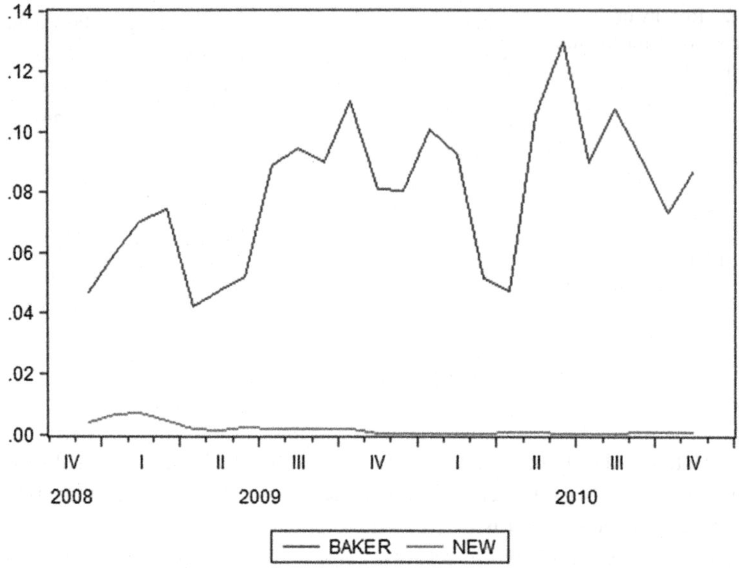

Fig. 1.5 Comparison chart of rolling *p*-values of the two indices

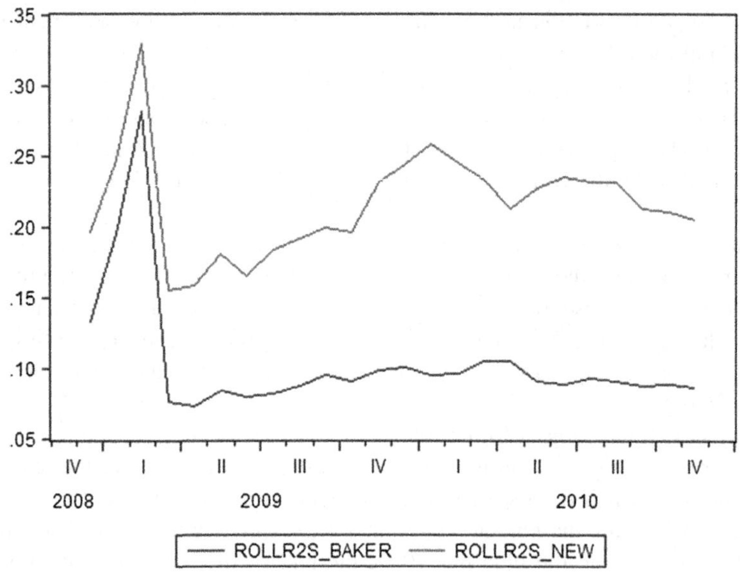

Fig. 1.6 Comparison chart of rolling *R*-squared of the two indices

Summarizing, it is obvious from the results that our new sentiment index created from the classification procedure, outperforms the sentiment index created by Baker and Wurgler, for the timeframe examined.

1.6 Conclusions and Future Work

In the recent years, investor sentiment has become less of an abstract idea and more of a precise measure helpful in both explaining and forecasting stock returns. This paper proposed a new and direct measure of investor's sentiment using newspaper articles, mainly provided by The New York Times. The sentiment index is created using a hybrid method that combines a Naïve Bayes classifier and the n-gram probabilistic language model.

First, a large amount of data for ten big companies and indices, which are being traded in the stock exchange, is collected from The New York Times web interface and stored in the NoSQL database MongoDB. Secondly, all articles downloaded are processed and manipulated in order to transform into a word sequence, which has reduced noise and is capable of being passed to the dictionaries and match any positive and negative word occurrences, so that a sentiment score can be extracted. After all articles get a score, three lists are created that contain neutral, most positive and most negative articles, and which will be used to train the classifier.

Once the classifier is created, we can pass unseen newspaper articles to it, in order to classify them as positive, negative or neutral and create a sentiment index for the company or index examined. The tool developed in this work is able to extract a sentiment score for daily, monthly and annual time frames, so it can match most of investors' trading strategies. It can also be extended to any company or index one might be interested in and for any time period.

The experiments performed in this work were based on 10-fold cross-validation, which is a resampling validation technique for assessing how the results of a statistical analysis will generalize to an independent new data set. With the cross-validation procedure, the average performance of the classification task is recorded and the misclassification error is measured. Besides performance estimation, other experiments deal with detecting the forecasting property of the new sentiment index created by the sentiment analysis for a company or index.

In a sample of S&P 500 index from 2004 to 2014 on monthly basis, it is shown that newspaper articles are correlated with the closing prices and the returns. In addition, the new sentiment index created is compared with the sentiment index created by Baker and Wurgler, and it is proven that for the existing time frame, the new index outperforms Baker and Wurgler's index, in terms of predicting returns.

Future research could extend the new index and review its accuracy for future returns. Also, it could be interesting for future work to expand the index backwards to earlier dates and review its forecasting ability and, also, compare it with Baker and Wurgler's index. Furthermore, a good and useful idea would be to store newspaper articles for other companies and indices one may be interested in and apply the tool created in this work to test its effectiveness for both older data and, also, for real-time data to examine the performance on news that pop up instantaneously.

Acknowledgments We would like to give special thanks to Dr. Theologos Dergiades, Academic Associate of the School of Science & Technology at the International Hellenic University, Thessaloniki, Greece, for his advice and assistance whenever it was needed.

References

1. Antweiler, W., Frank, M.: Is all that talk just noise? The information content of Internet stock message boards. J. Finance **59**(3), 1259–1294 (2004)
2. Baker, M., Wurgler, J.: Investor sentiment and the cross-section of stock returns. J. Finance **61** (4), 1645–1680 (2006)
3. Barber, B.M., Odean, T., Zhu, N.: Do retail trades move markets? Rev. Finan. Stud. **22**, 151–186 (2009)
4. Bollen, J., Mao, H., Zeng, X.: Twitter mood predicts the stock market. J. Comput. Sci. **2**(1), 1–8 (2010)
5. Bram, J., Ludvigson, S.C.: Does consumer confidence forecast household expenditure? A sentiment index horse race. Econ. Policy Rev. **4**(2) (1998)
6. Das, S., Chen, M.: Yahoo! for Amazon: sentiment extraction from small talk on the Web. Manag. Sci. **53**(9), 1375–1388 (2007)
7. Dergiades, T.: Do investors' sentiment dynamics affect stock returns? Evidence from the US economy. Econ. Lett. **116**(3), 404–407 (2012). ISSN 0165-1765, http://dx.doi.org/10.1016/j.econlet.2012.04.018
8. Dickinson, B., Hu, W.: Sentiment analysis of investor opinions on twitter. Soc. Network **4**, 62–71 (2015)
9. Fang, H., Tao, T., Zhai, Ch. X.: A formal study of information retrieval heuristics. In: Proceedings of the 27th Annual International ACM SIGIR Conference on Research and Development in Information Retrieval (SIGIR '04). ACM, New York, NY, USA, pp. 49–56 (2004). doi:http://dx.doi.org/10.1145/1008992.1009004
10. Garcia, D.: Sentiment during recessions. J. Finance **68**(3), 1267–1299 (2013)
11. Iyengar, S.: Is Anyone Responsible? How Television Frames Political Issues. University of Chicago Press, Chicago (1991)
12. Klemola, A., Nikkinen, J., Peltomäki, J.: Investor Sentiment in the Stock Market Inferred from Google Search Volumes (2010)
13. McDonald, B.: Bill McDonald's Word Lists Page, University of Notre Dame (2013). http://www3.nd.edu/~mcdonald/Word_Lists.html
14. Pang, B., Lee, L.: Opinion Mining and Sentiment Analysis (2008)
15. Plous S (1993) The Psychology of Judgment and Decision Making. McGraw-Hill. ISBN 978-0-07-050477-6
16. Price, V., Tewksbury, D.: News Values and Public Opinion: A Theoretical Account of Media Priming and Framing. In: Barett, G.A., Boster, F.J. (eds.) Progress in Communication Sciences: Advances in Persuasion, vol. 13, pp. 173–212. Ablex, Greenwich, CT (1997)
17. Price, V., Tewksbury, D., Powers, E.: Switching trains of thought: the impact of news frames on readers' cognitive responses. Commun. Res. **24**(5), 481–506 (1997)
18. Qiu, L., Welch, I.: Investor Sentiment Measures. Brown University and NBER (2006)
19. Schmeling, M.: Institutional and individual sentiment: smart money and noise trader risk? **23**, 127–145 (2007)
20. Sehgal, V., Song, C.: SOPS: Stock Prediction using Web Sentiment. In: Seventh IEEE International Conference on Data Mining—Workshops, pp. 21–26 (2009)
21. Shiller, R.J.: Irrational Exuberance, 2nd edn. Princeton University Press, Princeton, New Jersey (2005)
22. Shleifer, A., Summers, L.H.: The noise trader approach to finance. J. Econ. Perspect. **4**, 19–33 (1990)

23. Spärck Jones, K.: A statistical interpretation of term specificity and its application in retrieval. J. Documentation **28**, 11–21 (1972). doi:10.1108/eb026526
24. Statman, M.: Normal investors, then and now. CFA Institute, Finan. Anal. J. **61**(2), 31–37 (2005)
25. Tumarkin, R., Whitelaw, R.F.: News or Noise? Internet Message Board Activity and Stock Prices (2000)
26. Tversky, A., Kahneman, D.: The Framing of decisions and the psychology of choice. Science **211**(4481), 453–458 (1981). doi:10.1126/science.7455683
27. Vliegenthart, R., Schuck, A.R.T., Boomgaarden, H.G., De Vreese, C.H.: News coverage and support for european integration, 1990–2006. Int. J. Public Opin. Res. **20**(4), 415–436 (2008)
28. Wolpert, D.H.: Stacked generalization. Neural Networks **5**, 241 (1992)

Chapter 2
On the Effect of Adding Nodes to TSP Instances: An Empirical Analysis

Gloria Cerasela Crişan, Elena Nechita and Vasile Palade

Abstract Our human society is experiencing complex problems nowadays, which require large amounts of computing resources, fast algorithms and efficient implementations. These real-world problems generate new instances for the classical, academic problems as well as new data collections that can be used for assessing the available solving packages. This paper focuses on the Traveling Salesman Problem, which is one of the most studied combinatorial optimization problems, with many variants and broad applications. In order to allow a smooth integration with the current Geographic Information Systems (GIS) technologies, the instances described in this work are specified by geographic coordinates, and they use the orthodromic distance. A sequence of similar instances is defined, and the characteristics of the state-of-the-art exact solver results on these instances are presented and discussed.

Keywords Combinatorial optimization · Traveling Salesman Problem · Exact algorithms · Metaheuristics · Orthodromic distance

2.1 Introduction

The Traveling Salesman Problem (TSP) has been intensively studied by researchers all over the world. A search process over the Web of Science citation indexing service, maintained by Thomson Reuters, performed for the period 2000–2015 with the keywords "Traveling Salesman Problem", returns more than 7000 entries.

G.C. Crişan (✉) · E. Nechita
Vasile Alecsandri University of Bacău, Bacău, Romania
e-mail: ceraselacrisan@ub.ro

E. Nechita
e-mail: enechita@ub.ro

V. Palade
Coventry University, Coventry, UK
e-mail: vasile.palade@coventry.ac.uk

I. Hatzilygeroudis et al. (eds.), *Advances in Combining Intelligent Methods*,
Intelligent Systems Reference Library 116, DOI 10.1007/978-3-319-46200-4_2

However, the work developed by the scientific community, related to the TSP and its variants, is definitely much more extensive. Discussions on the TSP are present in many processes, starting with the educational ones, offered by the higher education institutions that include Computer Science degrees in their curricula and ending with the research developed within the most famous world's institutes in Artificial Intelligence.

The attention given to this combinatorial optimization problem comes from its significance for both theoretical advances and real-world applications. In addition, the TSP is one of the standard test problems used in the performance analysis of discrete optimization algorithms.

The goal of this paper is to empirically study the behavior of an exact TSP solver on a set of very close real-world instances. This work considers that two TSP instances are similar if one is obtained by adding a new node to the other, together with the costs of the corresponding generated edges. We used real data, downloaded from free online repositories. In order to comply with the large geographic area where data are sampled from, we used the orthodromic (great-circle) distances as travel costs. This research is a combination of approaches, since it uses the modern GIS format in order to define a structured collection of TSP instances. This set is solved with a state-of-the-art academic exact solver, and the results are interpreted using the re-optimization framework. Moreover, the solver uses a combination of techniques.

The next section presents a brief overview of several TSP variants and approaches. Section 2.3 describes the design and the implementation of an experiment for studying the influence of adding one single node into a medium-sized TSP instance. Section 2.4 contains the results of our computational study and their interpretations. Section 2.5 summarizes the conclusions of this research.

2.2 TSP—The Problem and Its Variants

The *general TSP* considers a set of n vertices and the nonnegative pairwise costs associated to the edges between those vertices. The objective of the problem is to find the minimum cost tour passing exactly once through each vertex. Formally, the TSP is defined as follows.

Let $G = (V, A)$ be a complete graph, with $V = \{v_1, \ldots, v_n\}$ the set of n vertices and $A = \{(v_i, v_j) \mid v_i, v_j \in V\}$ the set of arcs. Let $C = (c_{ij})$ be the cost matrix associated with A, with c_{ij} being the cost of (v_i, v_j). The TSP consists in determining the least cost Hamiltonian tour of G [1].

It is worth noting that the general TSP sets up no restrictions over the cost function. Since the Hamiltonian circuit decision problem is NP-complete [2], the TSP is also NP-complete.

The TSP is called *symmetric* if $c_{ij} = c_{ji}$ for all $v_i, v_j \in V$, otherwise it is called *asymmetric*. Therefore, the symmetric TSP is defined on a complete undirected graph, while the asymmetric TSP (ATSP) is defined on a complete directed one. C satisfies the triangle inequality if and only if $c_{ij} + c_{jk} \geq c_{ik}$ for all $v_i, v_j, v_k \in V$. Hereinafter, if not specified otherwise, the TSP denotes the symmetric TSP.

The *metric TSP* is the TSP whose vertices lie in a metric space (all edge costs are symmetric and fulfil the triangle inequality). The *Euclidean TSP* is a subcase of the metric TSP. In this case, the vertices are placed in R^d, $d \geq 2$ and the cost is defined using the l_d norm [3]. The most addressed variant of the Euclidean TSP is that for $d = 2$, due to its interpretations and applicability in real-life situations.

The *2D Euclidean TSP* considers a set of vertices which correspond to n locations in R^2. The objective in this common variant is to find the shortest tour through all locations. The Vehicle Routing Problem (VRP) also derives from the 2D Euclidean TSP [4].

Complexity studies on the TSP [5, 6] show that the problem is computationally difficult. The exact optimization of the general TSP is NP-hard [7] and so is the Euclidean TSP [8]. Nevertheless, in time, researchers have studied significant variants and instances, which arise in conjunction with the applications that can be modelled with the TSP as mathematical support. For many of the instances, the hardness results may not necessarily apply [6]. The following subsections present several TSP variants and considerations related to their complexity.

2.2.1 The Traveling Salesman Problem—Variants and Their Complexity

Applegate et al. [9] provide an extensive view over the advances in the TSP, until 2011. Our survey is conducted by the need of approaching the particularly generated TSP instances, in connection with the possibility to reuse the knowledge about the optimum results already determined for previously studied instances. Nowadays, new TSP instances may arise from various modern applications. Industry, Logistics, Transportations face specific scenarios and need specific models and simulations. Uncertainty, randomness and technological advances (such as the use of drones, reminded below) made us look towards very new versions of the TSP.

The Probabilistic Traveling Salesman Problem (PTSP) is a more difficult variant of the classical TSP. It was introduced by Jaillet [10] and came from the necessity to adopt a model that takes into account the random real life phenomena [11]: for many delivery companies that perform pickup and delivery, not all the customers require daily visits. In the PTSP, a visiting probability between 0 and 1 is assigned to each vertex. On a daily basis, according to its visiting probabilities, each vertex requests a visit or announces the visit skipping. The PTSP solution is a complete a priori tour of minimal expected length, which gives the order for the nodes that requested a visit on a certain day. The other nodes are skipped [12]. The PTSP formulation, given in [10], is the following:

Let T be an a priori PTSP tour through n vertices of a given graph G, where each vertex i has a visiting probability p_i independently of the others. Let d_{ij} be the distance between i and j. Without any loss of generality, we may assume that the a priori tour T is (1, 2, …, n, 1). The problem is to find that complete a priori tour which minimizes the expected length $E[L_T]$:

$$E[L_T] = \sum_{i=1}^{n} \sum_{j=i+1}^{n} d_{ij} p_i p_j \prod_{k=i+1}^{j-1} (1 - p_k) + \sum_{i=1}^{n} \sum_{j=1}^{i-1} d_{ji} p_i p_j \prod_{k=i+1}^{n} (1 - p_k) \prod_{l=1}^{j-1} (1 - p_l)$$

Obviously, the TSP is a special case of the PTSP, in which all the n vertices have $p_i = 1$. For instances with up to 50 vertices, branch-and-bound algorithms [13] and exact branch-and-cut algorithms [14] have been proposed to find the PTSP optimal solutions. However, for larger instances, most approaches to the PTSP are also based on heuristics and metaheuristics.

The Traveling Salesman Problem with Drone (TSP-D) is a new variant of the TSP [15]. Over the last years, many engineers have studied the possibility of using drones in combination with vehicles, to support deliveries and make them more cost-effective [16]. The idea involves both assignment decisions and routing decisions and it has generated the TSP-D.

Formally, the TSP-D is modeled in a graph $G = (V, A)$, where the node v_0 represents a depot and the n nodes $v_1, \ldots, v_n$ are the customer locations. The edges $e_{ij} = (v_i, v_j)$ connect the nodes v_i, v_j and c_{ij} is the driving time that a vehicle needs to commute from v_i to v_j or vice versa (therefore, the triangle inequality holds). The objective of the TSP-D is to find the shortest tour to serve all customers either by vehicle or by drone, in terms of time [15].

The TSP-D is related to the Covering Salesman Problem (CSP) introduced in 1989 by Current and Schilling [17], where the aim is to find the shortest tour of a subset of given nodes, while every node not belonging to the tour is within a predefined covering distance of a node on the tour. The TSP-D differs from CSP because the nodes outside the tour have to be visited by drones, and these have to be synchronized with the vehicles. This feature lines up the TSP-D with the class of the Vehicle Routing Problems (VRPs) [18] and also with the Truck and Trailer Routing Problem (TTRP) [19]. Moreover, the TSP-D is similar to the so-called "Flying Sidekick Traveling Salesman Problem", proposed in 2015 by Murray and Chu [20].

The TSP-D is also NP-hard. Under the assumptions that (a) the drones return to the truck after each delivery and (b) the pickups from the truck always take place at a customer location or at the depot, a solution would be a pair of tours (R, D), where R is a vehicle route from v_0 to v_0, together with a drone route D which includes all the customers that are visited by both truck and drone [15].

Introduced in 2004, the **one-commodity pickup-and-delivery Traveling Salesman Problem** (1-PDTSP) is a TSP plus the following additional constraints [21]: one specific city is the depot for a vehicle (with a fixed upper limit capacity) which visits the customers divided into two groups, according to the required service: delivery or pickup. Each delivery customer demands a given amount of a product, while each pickup customer provides a given amount of that product. Any amount collected from the pickup clients can be transferred to the delivery customers. The solution is a Hamiltonian tour through the cities, without ever exceeding the vehicle capacity. Several studies show that finding feasible 1-PDTSP solutions is much more complicated than finding good heuristic solutions for TSP [21, 22]. Combined approaches, involving the TSP heuristics and the branch-and-cut algorithms provide good solutions for the large instances, and also for the classical TSP with pickup and delivery (TSPPD) [23].

2.2.2 The Traveling Salesman Problem—Approaches

Apart from the applications directly related to (vehicle) routing, literature describes situations that can be modeled by the TSP, coming from various domains: computer wiring and dashboard design, hole punching and wallpaper cutting, applications in crystallography and in polyhedral theory [24]. Most of them are large scale applications and cannot be tackled using exact algorithms. Instead, heuristics are used to provide solutions, which can be considered very good in terms of computing time and/or computer resources. A short review over the most important exact and approximate algorithms is given in the following.

The Exact Methods for the TSP are connected to the developments in the field of the Integer Linear Programming (ILP). In this framework, the TSP formulation of Dantzig, Fulkerson and Johnson [25], further on referred as with DFJ, comes from 1954.

Let x_{ij} be a binary variable associated to the arc (v_i, v_j), with $x_{ij} = 1$ if and only if (v_i, v_j) belongs to the optimal solution. The DFJ formulation is as follows:

Minimize

$$\sum_{\substack{i,j=1\\ i\neq j}}^{n} c_{ij}x_{ij} \tag{2.1}$$

subject to

$$\sum_{j=1}^{n} x_{ij} = 1, \quad i = 1, \ldots, n \tag{2.2}$$

$$\sum_{i=1}^{n} x_{ij} = 1, \quad j = 1, \ldots, n \tag{2.3}$$

$$\sum_{i,j\in S} x_{ij} \leq |S| - 1, \quad S \subset V, \quad 2 \leq |S| \leq n-2 \tag{2.4}$$

$$x_{ij} \in \{0, 1\}, \quad i,j = 1, \ldots, n, \quad i \neq j \tag{2.5}$$

The objective function (2.1) describes the cost of the optimal tour. The constraints (2.2) and (2.3) specify that every vertex is entered exactly once and left exactly once (degree constraints). The forbiddance of tours on the subsets S of less than n vertices is given in (2.4) (subtour elimination constraints); as well, S cannot be disconnected (there must be at least one arc pointing from S to its complement). The binary conditions on the variables are described by (2.5). Under $n(n-1)$ variables, $2n$ degree constraints and 2^n-2n-2 subtour elimination constraints, the DFJ cannot be solved by ILP code even for moderate values of n. In order to approach the larger problems in the context of mathematical programming, alternative formulations have been proposed by relaxing the constraints (2.4), thus resulting assignment problems (AP) which can be solved in $O(n^3)$ time [26].

Based on the AP relaxation, several *branch-and-bound* (B&B) algorithms have been developed since 1958. Laporte [26] presents those proposed by Carpaneto and Toth (1980), [27] Balas and Cristophides (1981), [28] Miller and Pekny (1991) [29] among the best available, which provided the optimal solution for randomly generated problems with thousands of nodes (and for some real problems as well) in a reasonable CPU time. Various lower bound procedures were embedded in the B&B algorithms in order to improve their performances, but literature describes that their success depends essentially on the type of the problem to be solved. The symmetric TSPs are better handled by such algorithms [30].

The *branch-and-cut* (B&C) algorithms calculate series of increasing lower and decreasing upper bounds of the optimal solution. The upper bounds are given by heuristic algorithms, while for the lower bounds, the algorithms use a polyhedral cutting-plane procedure over the system of linear inequalities [31]. As a combination between B&B and a cutting procedure, B&C is usually faster than B&B alone. When the cutting-plane procedure does not terminate with an optimal solution, the algorithm uses a tree-search strategy that produces cuts after branching. A B&C algorithm eventually returns the optimal solution when the upper and lower bounds coincide. If this situation is not reached, the quality of the feasible solution can be precisely estimated [32]. The research performed by Jünger et al. [33] extensively describes the branch-and-cut implementation details.

Solvers and other software packages for the TSP

A *solver software* takes an instance of a problem as input, applies one or more methods and provides the result. A more extensive concept is that of the *modeling software*, which provides an environment for formulating, solving and analysing a problem. Modeling software implies at least one solver, but it usually offers several solvers. The following paragraphs provide information related to such products that allow developing research on the TSP. In [34], the reader can find a list of general-purpose software and libraries (released until 2007) for approaching the TSP applications. A survey realized by OR/MS Today Magazine [35] in June 2015 presents a significant list of Linear Programming software packages, together with the corresponding types (solver/modeling environment/integrated solver and modeling environment), supported platforms, vendors and pricing information.

COIN-OR (Common Optimization INterface for Operations Research) Repository [36] points to a large collection of library of interoperable software tools for building optimization codes, and it also includes several stand-alone packages. It was developed within the COIN-OR (COmputational INfrastructure for Operations Research) Project [37], an initiative promoting the use of interoperable, open-source software for Operations Research. The branch, cut and price library (written in C++) gives researchers the framework for customizing the LP solver according to their needs. COIN-OR also includes an object-oriented Tabu Search framework.

BOB (BOB++) [38] is a general-purpose software library that implements solvers for combinatorial optimization problems on parallel and sequential machines. The methods used are branch-and-bound and divide-and-conquer.

ABACUS (A Branch-And-Cut System) [33, 39] is an open source C++ system providing a framework for the implementation of the B&B and B&C algorithms.

The Concorde TSP Solver package, proposed in 2001, is still one of the fastest TSP exact solvers [40]. It uses the branch-and-cut and the Chained Lin-Kernigan implementations. It is freely available at [41].

Heuristics for the TSP

The heuristics for the TSP can be broadly classified in three classes: *Tour construction procedures*, *Tour improvement procedures* and *Composite procedures*.

The heuristics of the first type gradually build a solution by adding a new vertex at each step. The *Nearest neighbour heuristic* and the *insertion procedures* fall into this class. The insertion procedures use various criteria [1].

The heuristics of the second type use *improvement procedures* that start with a feasible solution and perform various exchanges upon it. The procedures proposed by Lin [42], Lin and Kernigan [43], and Or [44] are known as classical improvement procedures, lying behind the numerous attempts of improvements and hybrid versions.

The *r-opt* algorithm proposed by Lin [42] generalizes the systematic *2-opt* method suggested by Croes [45] in 1958. At a given iteration, r arcs (or edges, if the problem is symmetric) are removed from the current tour and all the possible reinsertions are attempted. The best is implemented and the operation is repeated until no further improvement is possible [46]. In 1973, Lin and Kernigan [43] developed the idea by allowing r to vary during the search, thus introducing a dynamic *r-opt* heuristic. Numerous heuristics for the TSP implemented this approach and proved to be very efficient. The *Or-opt*, introduced by Or in 1976 [44], attempts to improve the current tour by first moving a chain of three consecutive vertices in a different location, until no further improvement can be obtained. The process is repeated with chains of two consecutive vertices, and then with single vertices.

Although the Lin-Kernigan procedure is recognized as (computationally) effective, sometimes simpler heuristics such as *2-opt*, *3-opt* and *r-opt* are easier to implement and provide good performances. In [47], the authors develop and compare a number of *Or-opt* variants and (*2-opt*, *Or-opt*) hybrid procedures for the symmetric TSP.

The third class of heuristics includes two-phase construction procedures that combine procedures of the two previous types. The *CCAO* (*Convex hull, Cheapest insertion, Angle selection and Or-opt*) *heuristic* proposed by Golden and Stewart [48] falls into this class.

Metaheuristics for the TSP

The broad area of metaheuristics that proved to be very efficient for the TSP and its variants includes *Single-solution methods* and *Population-based methods*.

The single-solution techniques center on improving one candidate solution. Simulated Annealing uses a decreasing probability for accepting a worse solution, in order to expand the search and to escape from the local optimum [49]. Tabu Search uses memory structures to avoid considering an already visited solution and also randomly accepts worse solutions [50]. The Variable Neighborhood Search employs a set of neighborhoods and systematically changes them when a local optimum is reached [51].

The population-based methods include classical approaches, such as Genetic Algorithms (GA), Particle Swarm Optimization (PSO) and Ant Colony Optimization (ACO), as well as very novel meta-heuristics. Among these, the Imperialist Competitive Algorithm (ICA), the Artificial Bee Colony (ABC) and the Firefly Algorithm (FA) have already been implemented and applied in various contexts. While ICA [52] simulates the imperialist competition of the country, the nature-inspired methods copy the efficient social behaviour of populations such as ants, fireflies or bees, and different types of communication that these insects exhibit. Ants use pheromones for finding the shortest paths to the food source [53, 54], fireflies use bioluminescent communication [55], and the honey bee swarm divides its tasks between the cooperating groups of employed bees, onlookers and scouts [56]. In all these cases, various principles of communication and collaboration of such simple agents emerge intelligent behaviour at the community level [57–59].

The Bat Algorithm (BA) is a population-based metaheuristic introduced in 2010. The real bats can find their prey and differentiate among different kinds of insects in complete darkness [60]. This characteristic, based on echolocation, has been modeled and led to the BA, firstly proposed by Yang [61] for solving continuous problems. The Fuzzy Logic BA (FLBA) [62] and the Chaotic BA (CBA) [63] are among the versions [64] further developed for the BA, with very interesting applications such as in the study of the dynamic systems. As reviewed in [65], numerous versions of the Binary BA (BBA) were designed for addressing very diverse discrete optimization problems such as selection, planning, and flow shop scheduling. In 2015, an improved version (IBA) was proved [65] to perform with better performances, both for the TSP and for the ATSP.

The TSP Inexact Solvers

Based on the numerous available metaheuristics, the inexact solvers have been developed. The solver LKH [66] is a stochastic local search algorithm, based on the

Lin-Kernighan procedure. This optimization procedure involves swapping sets of *k* edges to generate feasible, possibly better solutions. It generalizes the *2*-opt and *3*-opt basic moves by dynamically deciding the value of the parameter *k* and seeking for better *k*-opt moves [67]. The LKH was known as the best inexact solver since 2000 and remained so until the development of EAX.

EAX has been recently introduced by Nagata and Kobayashi [68]. It is an evolutionary algorithm that uses 2-opt local search (for the initial population), a Tabu Search procedure (to generate offspring from high-quality parent solutions) and—as a specific feature—an edge assembly crossover procedure (providing very good quality tours by combining two parent tours with a small number of new, short edges). Some experiments [68] ran over commonly studied the Euclidean TSP instances provided, in some cases, with better results than those returned by the LKH.

Encouraging results were published in 2015 by Kotthoff et al. [69]. A series of empirical investigations with LKH and EAX witnessed performance improvements through the algorithm selection techniques [69], showing that per-instance selection between the two solvers can be helpful especially for large the TSP instances.

2.2.3 Benchmarks for TSP

In order to assess the performance of the solutions proposed by researchers for the TSP, various benchmarks have been proposed and maintained, starting with the TSPLIB collection initiated in 1990 by Reinelt [70]. Test instances are now available for various geographic problems, for the national TSPs as well as for the collection designed for the industrial applications [71]. Other collections that are the most accessed by researchers are: the Algorithm Selection Benchmark Repository ASlib [72], the benchmark instances for the Traveling Salesman Problem with Time Windows [73]. For our computational study we have used the online geographical database from the Geonames repository [74].

2.3 Computational Experiment Methodology and Implementation

The common approach when facing a problem (or a given set of problem instances) is to design and to implement an algorithm for solving it (them) efficiently. The work presented in the following sections uses a specific solver, dedicated to a specific problem, and explores its behavior when solving a set of connected instances. The instance dimension, restrictions and structure can have an important effect on the resources needed by the solving application. Predicting the execution time and/or the optimum solution when solving a specific problem instance can be very important in operational decisions, for example in case of disasters or military attacks [75, 76].

Discovering difficult instances can be helpful for solvers' designers, who can continuously improve their applications to better treat these reluctant cases. Extracting features that influence the difficultness of an instance can orient the research in optimization through narrowing the investigated solution space and consequently in finding more quickly the optimum solution. Anticipating the solving time for an instance, when knowing the solving time for a close instance, can be a decisive factor in choosing between computerized industrial or financial support systems.

The theoretical concept of re-optimization introduced by Bőckenhauer [77] is very close to this empiric investigation. The basic idea of re-optimization is to use the knowledge gathered by solving the previous similar instances. The re-optimization investigations on the TSP consider that two instances are similar if they have the same dimension and their cost functions differ only on one edge. In our work, two instances are similar if one is obtained by adding one node into the other. This experiment is also in line with other early research [78, 79].

Our computational experiment, which is in line with other early research, investigates the behavior of the NEOS server [80] for Concorde when solving a sequence of similar TSP instances. Taking into account that the classical benchmarks earlier mentioned consist of different classes of quasi-independent instances, we constructed a sequence of 31 highly correlated instances.

We started by downloading the data from [81]. This file contains all the world localities with more than 15,000 inhabitants. Each city is specified by the geographic coordinates in decimal format (one real number for the latitude and one real number for the longitude, with positive values for North and East, respectively). This format follows the current ISO recommendation: "For computer data interchange of latitude and longitude, ISO 6709:2008 generally suggests that decimal degrees be used" [82].

When extracting only the European cities and sorting them in a decreasing order by population, we obtained a list with 5978 cities. We decided to derive a sequence where each instance differs from the precedent one by a single node. The instance *europe5000.tsp* has the first 5000 most populous European cities. We added 30 times the next node from the list with 5978 European cities and thus we obtained a sequence of 31 real-world TSP instances. The instances in this collection are specified by their geographic coordinates, since the orthodromic (great-circle) distance between two geo-points on the Earth surface is at the core of the current GIS technologies, which are becoming an essential part of our postindustrial world's digital infrastructure.

Since each instance adds just one node to the previous instance, we expected the following:

- the execution time to slightly differ between two neighbor instances, and
- the optimum tour for the next instance to be likely to simply connect the new node to the optimum tour of the current instance (as in Fig. 2.3a), where the new

Table 2.1 Optimum length (m), solving time (s), type of insertion, the new city and its country for each of the 31 TSP instances

# of nodes	Optimum length (m)	Total solving time (s)	Simple insertion (Yes/No)	Inserted city	Country
5,000	112,648,571	8,743	–	–	–
5,001	112,650,156	9,187	Yes	la Nucia	ES
5,002	112,817,600	6,909	Yes	Mo i Rana	NO
5,003	112,871,915	4,141	No	Forssa	FI
5,004	112,873,153	3,803	Yes	Zuerich (Kreis 11)/ Seebach	CH
5,005	112,903,052	5,893	No	Osuna	ES
5,006	112,911,768	7,912	Yes	Brixham	GB
5,007	112,912,015	7,735	Yes	Amorebieta	ES
5,008	112,912,507	6,746	Yes	Oria	ES
5,009	112,920,189	3,512	No	Reinheim	DE
5,010	112,920,864	5,897	Yes	Kristinehamn	SE
5,011	112,921,034	5,091	Yes	Libiaz	PL
5,012	112,949,261	6,701	Yes	Kukes	AL
5,013	112,949,777	8,158	Yes	Bastia Umbra	IT
5,014	112,952,180	9,377	Yes	Maesteg	GB
5,015	112,977,334	7,454	No	Acqui Terme	IT
5,016	112,983,928	7,917	No	Bunschoten	NL
5,017	112,983,944	8,367	Yes	Bilopillya	UA
5,018	112,990,645	10,905	Yes	Holzwickede	DE
5,019	112,999,351	9,979	Yes	Cercola	IT
5,020	113,008,809	12,251	No	Bohodukhiv	UA
5,021	113,008,928	9,419	Yes	Orsay	FR
5,022	–	–	–	Renens	CH
5,023	113,015,107	11,300	Yes, Yes	Brakel	DE
5,024	–	–	–	Saint-Amand-les-Eaux	FR
5,025	–	–	–	Teo	ES
5,026	113,024,494	10,573	No, Yes, Yes	Zubia	ES
5,027	113,032,048	10,314	Yes	Weesp	NL
5,028	113,061,960	9,961	No	Weissenburg in Bayern	DE
5,029	113,073,099	12,530	No	Palanga	LT
5,030	113,076,593	8,622	Yes	Moita	PT

node Brixham is connected by two red edges, and the blue edge between Paignton and Torquay is deleted.

The results presented in the next section show that both our suppositions are wrong.

2.4 Results and Discussion

In order to test these hypotheses, we called the solver Concorde with default parameters and we recorded for each instance: the execution time, the optimum length, and whether the optimum tour for the next instance differs by only two edges from the optimum tour for the previous one. The results are presented in Table 2.1.

Figure 2.1 shows the optimum tour for the *europe5000.tsp* instance. All the maps from this paper are drawn with the GPSVisualizer [83].

Figure 2.2 presents the 30 new added nodes. They are sampled all over Europe, close to the repartition of the initial 5000 nodes. The countries with the most cities in *europe5000.tsp* are: Germany (882 nodes), Great Britain (610 nodes), France (529 nodes) and Italy (468 nodes). There is no pair of consecutive inserted cities that are close to each other on the map.

Figure 2.3a illustrates the position of the 5006th node (Brixham) into the optimum tour: two red edges are inserted into the optimum tour for the *europe5005.tsp* instance, and the blue edge connecting Paignton with Torquay is deleted. We call

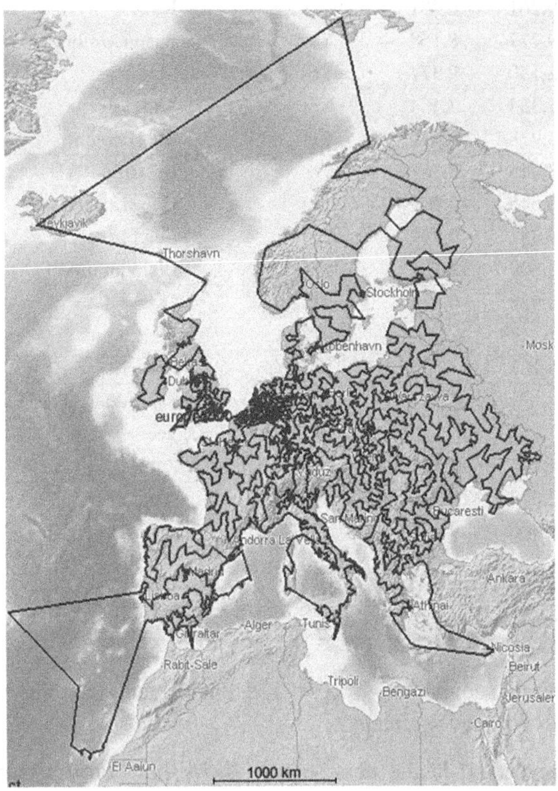

Fig. 2.1 Optimum tour for the *europe5000.tsp* instance

this situation a *simple insertion*. Figure 2.3b shows a more complicated effect of the 5009th node (Reinheim, in the bottom of the image) insertion: the optimum tour for the *europe5008.tsp* instance receives 11 red edges, loses 10 blue edges and becomes the optimum tour for the *europe5009.tsp* instance.

The administrators of the online Concorde solver kill the execution jobs after 4 h, if this is unsuccessful. The first unsolved instance in our experiment was *europe5022.tsp*. The application could not provide the optimum solution in the allocated time. It is interesting that *europe5023.tsp* was solved in 3 h and 9 min, and only two simple insertions into the optimum tour for the *europe5021.tsp* instance were needed (Fig. 2.4).

The next two instances are also difficult. The *europe5026.tsp* instance was solved in less than 4 h (actually in less than 3 h). Its optimum tour has two simple insertions (Fig. 2.5) into the previous, known optimum tour, but it also has a very large tour modification (Fig. 2.6). The optimum tour for *europe5026.tsp* also contains a major path of the same length (red path, central part of Fig. 2.6).

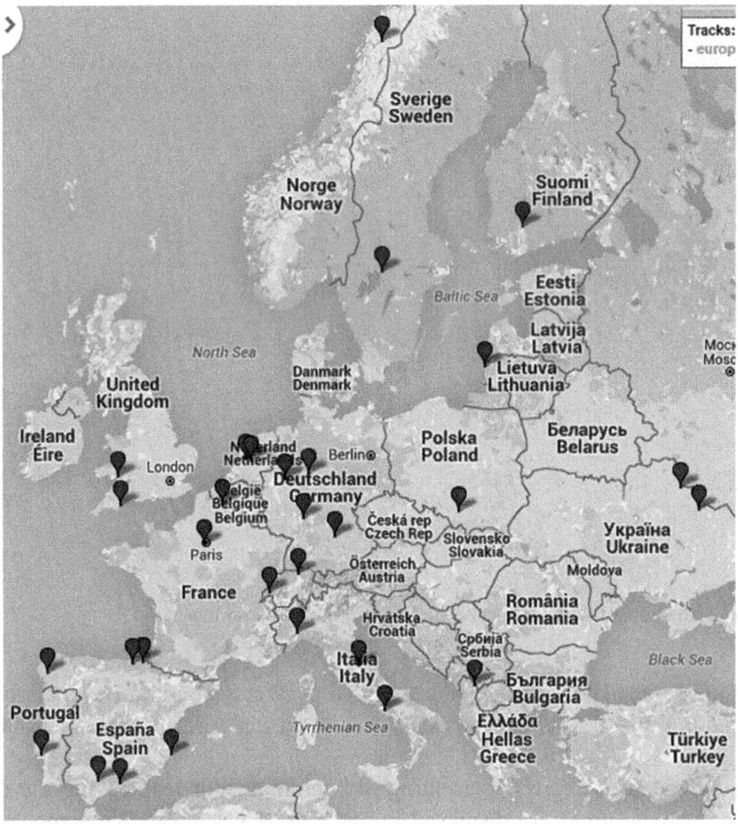

Fig. 2.2 The representation of the 30 nodes sequentially added

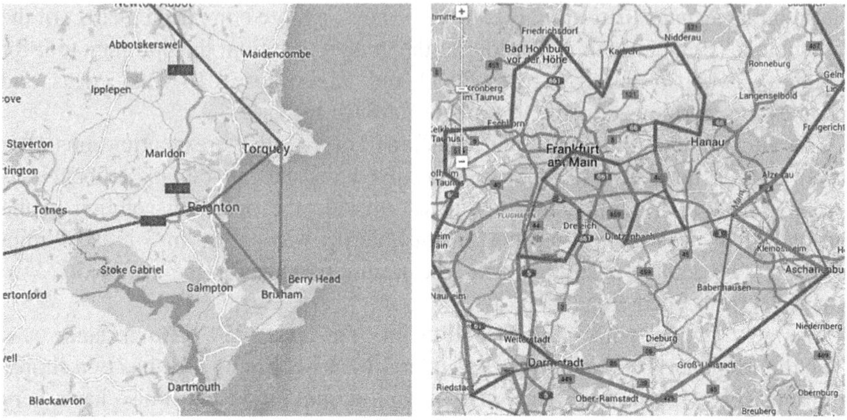

Fig. 2.3 **a** Optimum tour for the europe5006.tsp instance: simple insertion into the europe5005. tsp optimum tour (two new *red* edges instead of one *blue* edge). **b** Optimum tour for the europe5009.tsp instance: 11 new *red* edges replace 10 *blue* edges from the optimum tour for europe5008.tsp

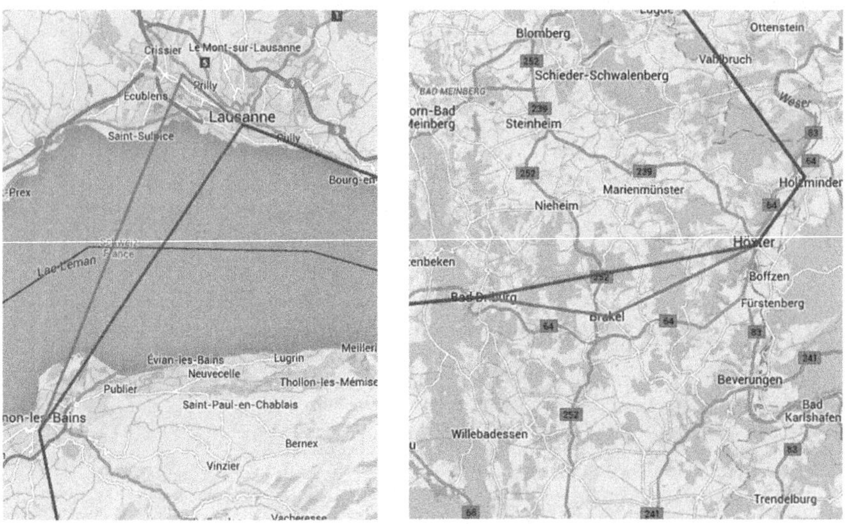

Fig. 2.4 Two simple insertions for the *europe5023.tsp* optimum tour

The most remarkable change was triggered by Weissenburg in Bayern, which is presented in Fig. 2.7. The new optimum tour covers an estimated area of 1,800,000 square km.

The independent charts representing the optimum tour lengths and the corresponding solving times are displayed in the Figs. 2.8 and 2.9. We computed the Pearson correlation coefficient for the optimum lengths and the solving time. We

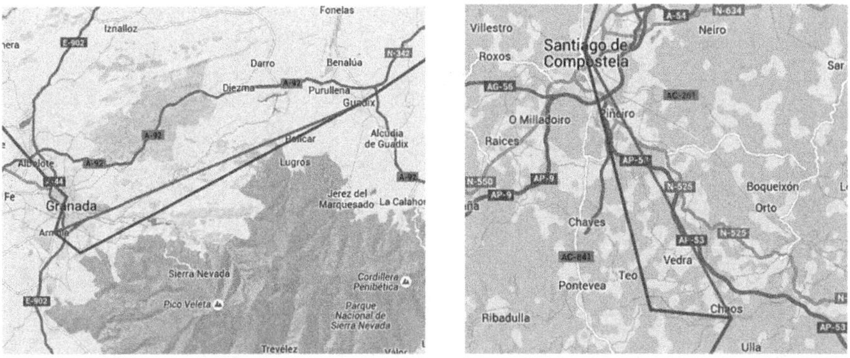

Fig. 2.5 Two simple insertions for the europe5026.tsp optimum tour, generated by the two Spanish cities

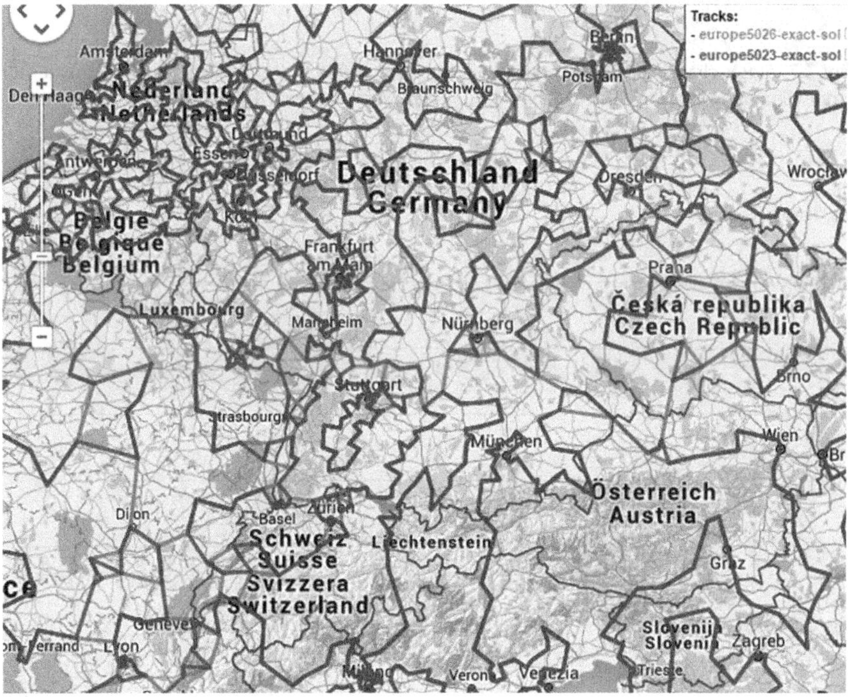

Fig. 2.6 Large change in the optimum tour for *europe5026.tsp*, generated by the French city (*upper left*, at the Belgium border), and a new path covering Central Europe

obtained the value r = –0.041789, which shows that the two samples do not correlate. This means that if the new city produces a small increase to the tour length, it does not necessarily generate an easier-to-solve instance, and vice versa.

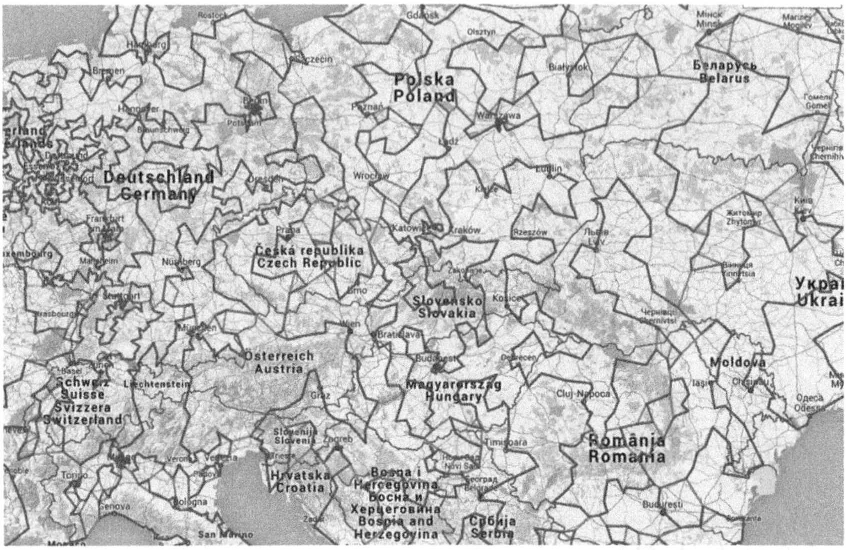

Fig. 2.7 Huge change in the optimum tour for *europe5028.tsp*, generated by the German city (from Ukraine to France, from Poland to Bulgaria, estimated area: 1,800,000 square km)

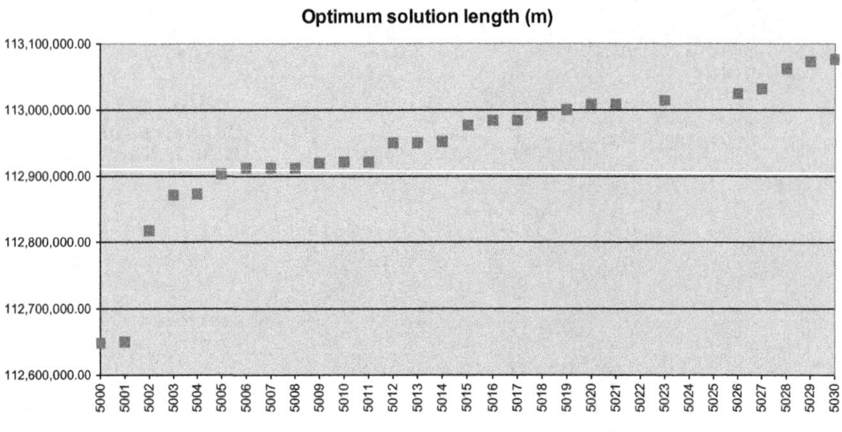

Fig. 2.8 The optimum solution lengths (m) for the TSP instances

For example, *europe5010.tsp* is only 675 m longer than the previous instance and it needed 38 more minutes. The instance *europe5002.tsp* is 167 km longer than *europe5001.tsp*, and it was solved in half an hour less.

According to the Fig. 2.9, we can notice the discontinuity of the solving time. For example, only 9 values lie in a 10 % distance from the previous one. Table 2.1 shows that in 21 cases simple insertions were made. In these cases, there is no pattern manifested in the solving time values. For example, the set *europe5011.tsp*–

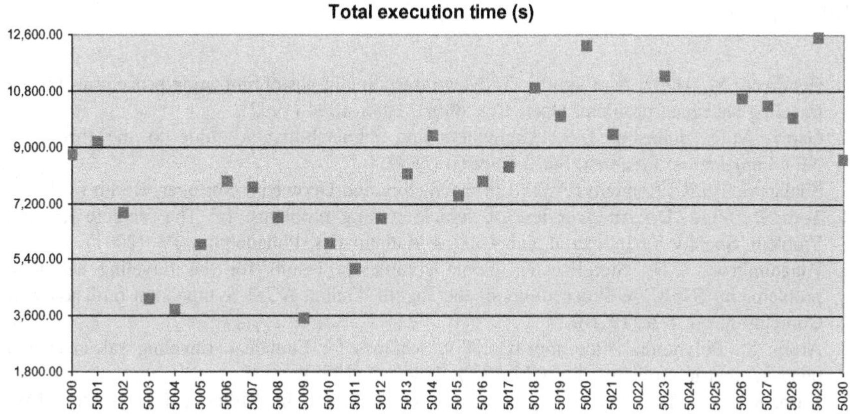

Fig. 2.9 The exact solving time (s) for the TSP instances

europe5014.tsp, whose instances have only simple insertions, also has solving times that differ each by more than 1000 s from the previous one. By exclusion, in 30 % of the cases, the effect of the new node was not minimal.

We can conclude that the sequence of the real-world TSP instances we used in our experiment manifest hard-to-predict characteristics of the optimum solutions. Moreover, the solving time was not continuous and the insertion of a new node into the optimum solution was not simple either, as frequent as we expected.

2.5 Conclusions and Future Work

This paper investigates the behavior of an exact TSP solver when it approaches similar instances. We used a set of 31 instances with medium size dimension, representing European localities. Each instance adds a new settlement to the previous one. The results of this empirical study show that no correlation can be highlighted between the lengths of the best solutions, the execution times and the complexity of the transformation of the previous best tour into the current best tour.

Future work has to explore the reasons of this behavior and the structure of the instances that are easier to re-optimize.

The collection of large area, medium dimension, highly correlated TSP instances we defined in this paper can be used to test logistic, emergency management or military decision support systems. Similar real-world, structured collections of instances can be defined by academics. As the current libraries of TSP instances are not correlated, this insight could open new research paths.

Acknowledgments G.C.C. and E.N. acknowledge the support of the project "Bacau and Lugano —Teaching Informatics for a Sustainable Society", co-financed by a grant from Switzerland through the Swiss Contribution to the enlarged European Union.

References

1. Gendreau, M., Hertz, A., Laporte, G.: New insertion and postoptimization procedures for the traveling salesman problem. Oper. Res. **40**(6), 1086–1094 (1992)
2. Garey, M.R., Johnson, D.S.: Computers and intractability: a guide to the theory of NP-completeness. Freeman, San Francisco (1979)
3. Shafarevich, I.R., Remizov, A.O.: Linear Algebra and Geometry. Springer, Berlin (2013)
4. Toth, P., Vigo, D.: An overview of vehicle routing problems. In: The Vehicle Routing Problem. Society for Industrial and Applied Mathematics, Philadelphia, PA (2001)
5. Papadimitrou, C.H., Steiglitz, K.: Some complexity results for the traveling salesman problem. In: STOC'76 Proceedings of the Eighth Annual ACM Symposium on Theory of Computing, pp. 1–9 (1976)
6. Arora, S.: Polynomial time approximation schemes for Euclidean traveling salesman and other geometric problems. J. ACM **45**(5), 753–782 (1998)
7. Karp, R.M.: Reducibility among combinatorial problems. In: Miller, R.E., Thatcher, J.W. (eds.) Complexity of Computer Computations, Advances in Computer Research, pp. 85–103. Plenum Press, (1972)
8. Papadimitrou, C.H.: Euclidean TSP is NP-complete. Theoret. Comput. Sci. **4**, 237–244 (1977)
9. Applegate, D.L., Bixby, R.E., Chvatal, V., Cook, W.J.: The Traveling Salesman Problem: A Computational Study. Princeton University Press, Princeton, NJ, USA (2011)
10. Jaillet, P.: Probabilistic Traveling Salesman Problems. PhD Thesis, MIT, Cambridge, MA, USA (1985)
11. Henchiri, A., Ballalouna, M., Khansaji, W.: Probabilistic traveling salesman problem: a survey. In: Position Paper of the 2014 Federated Conference on Computer Science and Information Systems, pp. 55–60 (2014)
12. Jaillet, P.: A priori solution of a traveling salesman Problem in which a random subset of the customers are visited. Oper. Res. **36**, 929–936 (1988)
13. Berman, O., Simchi-Levi, D.: Finding the optimal a priori tour and location of a traveling salesman with nonhomogenous customers. Transp. Sci. **22**(2), 148–154 (1988)
14. Laporte, G.: The traveling salesman problem: an overview of exact and approximate algorithms. Eur. J. Oper. Res. **59**, 231–247 (1992)
15. Agatz, N., Bouman, P., Scmidt, M.: Optimization approaches for the traveling salesman problem with drone. Technical Report, ERIM report series research in management (2015). http://repub.eur.nl/pub/78472
16. Popper, B.: UPS researching delivery drones that could compete with Amazon's Prime Air (2013). http://www.theverge.com/2013/12/3/5169878/ups-is-researching-its-own-delivery-drones-to-compete-with-amazons
17. Current, J.R., Schilling, D.A.: The covering salesman problem. Transp. Sci. **23**(3), 208–213 (1989)
18. Caric, T., Gold, H. (eds.): Vehicle Routing Problem. I-Tech Education and Publishing KG, Vienna (2008)
19. Derigs, U., Pullmann, M., Vogel, U.: Truck and trailer routing—problems, heuristics and computational experience. Comput. Oper. Res. **40**(2), 536–546 (2013)
20. Murray, C.C., Chu, A.G.: The flying sidekick traveling salesman problem: optimization of drone-assisted parcel delivery. Transp. Res. Part C: Emerg. Technol. **54**, 86–109 (2015)
21. Hernández-Pérez, H., Salazar-González, J.J.: Heuristics for the one-commodity pickup-and-delivery traveling salesman problem. Transp. Sci. **38**(2), 245–255 (2004)
22. Fischeti, M., Lodi, A.: Local branching. Math. Program. **98**, 23–47 (2003)
23. Berbeglia, G., Cordeau, J.-F., Laporte, G.: Dynamic pickup and delivery problems. Eur. J. Oper. Res. **202**(1), 8–15 (2010)
24. Reinelt, G.: The Traveling Salesman: Computational Solutions for TSP Applications. Lecture Notes in Computer Science. Springer, Berlin (1994)

25. Dantzig, G.B., Fulkerson, D.R., Johnson, S.M.: Solutions of a large-scale traveling salesman problem. Oper. Res. **2**, 393–410 (1954)
26. Laporte, G., Louveaux, F., Mercure, H.: A priori optimization of the probabilistic traveling salesman problem. Oper. Res. **42**(3), 543–549 (1994)
27. Carpaneto, G., Toth, P.: Some new branching and bounding criteria for the asymmetric travelling salesman problem. Manage. Sci. **26**, 736–743 (1980)
28. Balas, E., Christofides, N.: A restricted lagrangean approach to the traveling salesman problem. Math. Program. **21**, 19–46 (1981)
29. Miller, D.L., Pekny, J.F.: Exact solution of large asymmetric traveling salesman problems. Science **251**, 754–761 (1991)
30. Schrijver, A.: Combinatorial optimization: polyhedra and efficiency, vol. 1. Springer, Berlin (2003)
31. Padberg, M., Rinaldi, G.: A branch-and-cut algorithm for the resolution of large-scale symmetric traveling salesman problems. SIAM Rev. **33**(1), 60–100 (1991)
32. Mitchell, J.E.: Branch-and-cut algorithms for combinatorial optimization problems. Handbook of Applied Optimization. Oxford University Press, Oxford (2000)
33. Jünger, M., Reinelt, G., Thienel, S.: Provably good solutions for the traveling salesman problem. Zeitschrift für Operations Research **22**, 83–95 (1998)
34. Gutin, G., Punnen, A.P. (eds.): The Traveling Salesman Problem and Its Variations. Springer, New York (2007)
35. OR/MS Today magazine, Institute for Operations Research and the Management Sciences: 2015 Linear Programming Software Survey. http://www.orms-today.org/surveys/LP/LP-survey.html
36. COIN-OR resources. http://www.coin-or.org/projects/, http://www.coin-or.org/resources.html
37. COIN-OR Project. http://www.coin-or.org/
38. Galea, F., Le Cun, B.: Bob++: a Framework for exact combinatorial optimization methods on parallel machines. In: Proceedings of the 21st European Conference on Modelling and Simulation (2007)
39. ABACUS system. http://www.informatik.uni-koeln.de/abacus/index.html
40. Applegate, D.L., Bixby, R.E., Chvatal, V., Cook, W.J.: TSP cuts which do not conform to the template paradigm. Computational Combinatorial Optimization. Springer, Berlin (2001)
41. Concorde TSP solver. http://www.math.uwaterloo.ca/tsp/concorde/
42. Lin, S.: Computer solutions of the traveling salesman problem. Bell Syst. Tech. J. **44**, 2245–2269 (1965)
43. Lin, S., Kernighan, B.W.: An effective heuristic algorithm for the traveling salesman problem. Oper. Res. **21**, 972–989 (1973)
44. Or, I.: Traveling salesman-type combinatorial problems and their relation to the logistics of regional blood banking, PhD thesis, North-Western University, Evanston, IL (1976)
45. Croes, G.A.: A method for solving large scale symmetric traveling salesman problems to optimality. Oper. Res. **6**, 791–812 (1958)
46. Rosenkrantz, D.J., Stearns, R.E., Philip, I., Lewis, M.: An analysis of several heuristics for the traveling salesman problem. SIAM J. Comput. **6**(3), 563–581 (1977)
47. Babin, G., Deneault, S., Laporte, G.: Improvements of the Or-opt Heuristic for the Traveling Salesman Problem. GERARD—Group for Research in Decision Analysis. Montreal, Quebec, Canada (2005). https://blogue.hec.ca/permanent/babin/pub/Babi05a.pdf
48. Golden, B.L., Stewart, Jr.W.R.: Empirical analysis of heuristics. In: Hawler, E.L., Lenstra, J. K., Rinnouy Kan, A.H.G., Shmoys, D.B. (eds.) The Traveling Salesman Problem, pp. 207–249. Wiley, New York (1985)
49. Kirkpatrick, S., Gelatt Jr., C.D., Vecchi, M.P.: Optimization by simulated annealing. Science **220**(4598), 671–680 (1983)
50. Glover, F., Laguna, M.: Tabu Search. Kluwer Academic Publishers (1997)
51. Hansen, P., Mladenovic, N., Perez, J.A.M.: Variable neighbourhood search: methods and applications. Ann. Oper. Res. **175**, 367–407 (2010)

52. Ardalan, Z., Karimi, S., Poursabzi, O., Naderi, B.: A novel imperialist competitive algorithm for generalized traveling salesman problems. Appl. Soft Comput. **26**, 546–555 (2015)
53. Dorigo, M., Stützle, T.: Ant Colony Optimization. MIT Press, Cambridge (2004)
54. Crişan, G.C., Pintea, C.M., Pop, P.: On the resilience of an ant-based system in fuzzy environments. an empirical study. In: Proceedings of the 2014 IEEE International Conference on Fuzzy Systems, Beijing, China, pp. 2588–2593 (2014)
55. Jati, G.K., Suyanto, S.: Evolutionary discrete firefly algorithm for traveling salesman problem, ICAIS 2011. Lecture Notes in Artificial Intelligence (LNAI 6943), pp. 393–403 (2011)
56. Karaboga, D., Basturk, B.: On the performance of artificial bee colony (ABC) algorithm. Appl. Soft Comput. **8**(1), 687–697 (2008)
57. Iantovics, B., Chira, C., Dumitrescu, D.: Principles of Intelligent Agents. Casa Cărţii de Ştiinţă, Cluj-Napoca (2007)
58. Nechita, E., Muraru C.V., Talmaciu M.: Mechanisms in social insect societies and their use in optimization. a case study for trail laying behavior. In: Proceedings of the 1st International Conference Bio-Inspired Computational Methods Used for Solving Difficult Problems—Development of Intelligent and Complex Systems BICS'2008, Târgu Mureş, AIP Conference Proceedings, Melville, New York (2009)
59. Pintea, C.M.: Advances in Bio-inspired Computing for Combinatorial Optimization Problems. Springer, Berlin (2014)
60. Brigham, R.M., Kalko, K.V., Jones, G., Parsons, S., Limpens, H.J.G.A (Eds.): Bat echolocation research: tools, techniques and analysis, Austin, Texas (2002)
61. Yang, X.S.: A new meta-heuristic bat-inspired algorithm. In: Gonzales, J.R., Pelta, D.A., Cruz, C., Terrazas, G., Krasnogor, N. (eds.) Nature Inspired Cooperative Strategies for Optimization (NICSO 2010), pp. 65–74. Springer, Berlin (2010)
62. Khan, K., Nikov, A., Sahai, A.: A fuzzy bat clustering method for ergonomic screening of office workplaces. In: Third International Conference on Software, Services and Semantic Technologies, pp. 59–66. Springer (2011)
63. Lin, J.H., Chou, C.W., Yang, C.H., Tsai, H.L.: A chaotic Levy flight bat algorithm for parameter estimation in nonlinear dynamic biological systems. J. Comput. Inf. Technol. **2**(2), 56–63 (2012)
64. Yang, X.S., He, X.: Bat algorithm: literature review and applications. Int. J. Bio-Inspired Comput. **5**, 141–149 (2013)
65. Osaba, E., Yang, X.S., Diaz, F., Lopez-Garcia, P., Carballedo, R.: An improved discrete bat algorithm for symmetric and asymmetric traveling salesman problems, engineering applications of artificial intelligence (2015, in press)
66. Helsgaun, K.: General k-opt submoves for the Lin-Kernighan TSP heuristic. Math. Program. Comput. **1**(2–3), 119–163 (2009)
67. Helsgaun, K.: Solving the Bottleneck Traveling Salesman Problem Using the Lin-Kernigan-Helsgaun Algorithm. Technical Report, Computer Science, Roskilde University (2014)
68. Nagata, Y., Kobayashi, S.: A powerful genetic algorithm using edge assembly crossover for the traveling salesman problem. INFORMS J. Comput. **25**(2), 346–363 (2013)
69. Kotthoff, L., Kerschke, P., Hoos, H., Trautmann, H.: Improving the state of the art in inexact TSP solving using per-instance algorithm selection. Lecture Notes in Computer Science, vol. 8994, pp. 202–217. Springer (2015)
70. Library of various sample TSP and TSP-related instances. http://comopt.ifi.uni-heidelberg.de/software/TSPLIB95/
71. TSPLIB. http://www.math.waterloo.ca/tsp/index.htm
72. Algorithm Selection Library ASlib. http://www.coseal.net/aslib/
73. Benchmark instances for the Traveling Salesman Problem with Time Windows. http://lopez-ibanez.eu/tsptw-instances
74. Geonames repository. http://www.geonames.org
75. Crişan, G.C., Pintea, C.M., Chira, C.: Risk assessment for incoherent data. Environ. Eng. Manag. J. **11**(12), 2169–2174 (2012)

76. Nechita, E., Muraru, C.V., Talmaciu, M.: A Bayesian approach for the assessment of risk probability. Case Study Dig. Risk Probab. Environ. Eng. Manag. J. **11**(12), 2249–2256 (2012)
77. Bőckenhauer, H.J., Hromkovič, J., Mőmke, T., Widmaye, P.: On the Hardness of Reoptimization, SOFSEM 2008. LNCS, vol. 4910, pp. 50–65. Springer, Heidelberg (2008)
78. Papadimitriou, C.H., Steiglitz, K.: Some examples of difficult traveling salesman problems. Oper. Res. **26**(3), 434–443 (1978)
79. Ahammed, F., Moscato, P.: Evolving L-systems as an intelligent design approach to find classes of difficult-to-solve traveling salesman problem instances. In: Applications of Evolutionary Computation *EvoApplications 2011*: EvoCOMPLEX, EvoGAMES, EvoIASP, EvoINTELLIGENCE, EvoNUM, and EvoSTOC, Torino, Italy, April 27–29, 2011, Proceedings, Part I, pp. 1–11. Springer, Berlin (2011)
80. NEOS server. http://www.neos-server.org/neos/
81. World cities with 15,000 people or more. http://download.geonames.org/export/dump/
82. ISO 6709:2008, Standard representation of geographic point location by coordinates. http://www.iso.org/iso/iso_catalogue/catalogue_tc/catalogue_detail.htm?csnumber=39242
83. GPSVisualizer. http://www.gpsvisualizer.com

Chapter 3
Comparing Algorithmic Principles for Fuzzy Graph Communities over Neo4j

Georgios Drakopoulos, Andreas Kanavos, Christos Makris and Vasileios Megalooikonomou

Abstract Fuzzy graphs occur frequently in diverse fields such as computational neuroscience, social network analysis, devops, and information retrieval. This chapter covers an important class of fuzzy graphs where vertices are fixed whereas edges are fuzzy and exist according to a given or estimated probability distribution. Empirical evidence strongly suggests that, similarly to their deterministic counterparts, large fuzzy graphs of this type consist of recursively nested communities. The latter are closely linked to efficient local information dissemination and processing. Two community discovery algorithms, namely Fuzzy Walktrap and Fuzzy Newman-Girvan, based on different algorithm design principles are proposed and the performance of their Java implementation over Neo4j is experimentally assessed in terms of both total execution time and average graph cut cost on synthetic and real fuzzy graphs.

Keywords Community detection · Edge density · Fuzzy graphs · Higher order data · Kronecker model · Large graph analytics · Membership function · Newman-Girvan algorithm · Termination criteria · Walktrap algorithm

3.1 Introduction

Nowadays, Twitter is among the most popular microblogging services worldwide. Every day, a vast amount of tweet information is published by users to the public or to selected circles of their contacts. This information comes in the form of *tweets*,

G. Drakopoulos (✉) · A. Kanavos · C. Makris · V. Megalooikonomou
Computer Engineering and Informatics Department,
University of Patras, Patras 26500, Hellas
e-mail: drakop@ceid.upatras.gr

A. Kanavos
e-mail: kanavos@ceid.upatras.gr

C. Makris
e-mail: makri@ceid.upatras.gr

V. Megalooikonomou
e-mail: vasilis@ceid.upatras.gr

I. Hatzilygeroudis et al. (eds.), *Advances in Combining Intelligent Methods*,
Intelligent Systems Reference Library 116, DOI 10.1007/978-3-319-46200-4_3

i.e. public posts up to 140 characters each. Exceeding 241 million active users, 500 million tweets, and 2.1 billion searches per day, users and online marketers in particular have an actively informed audience to engage with. The rise of Twitter has completely changed end users, transforming them from simple passive information seekers and consumers to active producers, as it represents one of the most dynamic online datasets of user generated and distributed content.

The increasing popularity of social media, including Twitter which we consider in the present work, has gained in recent years huge research interest as well as new opportunities for studying the interactions of different groups of people. The rapid growth of Twitter in number of users, has redefined its status from a simple social media service to an important widely used social media, where companies have the opportunity to examine what customers say about their products and services.

Community detection tries to analyze a social network with the aim of finding groups of associated individuals in it. Analyzing the way that users formulate social communities, the determination of user behavior in each one of the communities as well as in the whole social network are fundamental aspects of social network analysis. More specifically, studying the community structure of a network leads in explaining social dynamics of interaction among groups of individuals and several research works point to this direction [6].

Vast amount of empirical evidence suggests that large scale graphs such as brain connectivity graphs, protein-to-protein interaction graphs, transportation networks, and the Web graph, strongly tend to exhibit modularity. In other words, they are composed of recursively built communities, a crucial factor for scaling property [3, 11].

Communities are highly connected vertex subsets which communicate with each other with few long distance edges. The communities account for the quick local information diffusion and processing, whereas the long distance edges serve as exchange points. Often large communities can be further subdivided into smaller communities. For instance, in social media an automotive group may be analyzed to an F1 group, a car group, and a motorcycle group depending on the particular interests of its users.

As a result of the importance of community discovery in developing large graph analytics, various algorithms have been developed. Two of the most prominent ones are the Newman-Girvan and the Walktrap algorithms. The former is deterministic and is based on local edge density, whereas the latter is heuristic and relies on the concept of an edge traversing random walker. These radically different approaches indicate the flexibility inherent in graph analytics as well as the multitude of the ways a graph can be interpreted.

Notice that this work is an extended version of [8]. Concretely, the primary contribution of this work is the development of the fuzzy versions of Walktrap and Newman-Girvan algorithms. The proposed versions have been based on the fuzzy graph model introduced among others in [25] and its properties. They have been applied to a synthetic graph obtained by the Kronecker model with various termination criteria. Moreover, the experiments are conducted in a more detailed way, by solidly demonstrating the algorithmic notions outlined in [8].

Table 3.1 Symbols used in this work

Symbol	Meaning
$\triangleq$	Definition or equality by definition
$\sim$	Distribution according to a density function
$\otimes$	Kronecker tensor product
$\mathcal{N}\left(\mu_0, \sigma_0^2\right)$	Gaussian distribution with μ_0 and σ_0^2
$\left\{s_1, s_2, \ldots, s_n\right\}$	Set with elements $s_1, s_2, \ldots, s_n$
$\mid S \mid$	Cardinality of set S
$\langle x_k \rangle$	Sequence of items x_k
$\left(e_1, \ldots, e_m\right)$	Path comprised of edges $e_1, \ldots e_m$
ζ	Graph diameter
$\deg\left(v_k\right)$	Degree of vertex v_k
$\mathcal{H}\left(s_1, \ldots, s_n\right)$	Harmonic mean of $s_1, \ldots, s_n$
τ_{S_1,S_2}	Tanimoto similarity coefficient for S_1 and S_2
γ	Algorithm termination criterion
η	Termination criterion change rate
K_n	Complete graph with n vertices and $\binom{n}{2}$ edges

The remaining of this work is structured as follows. Scientific literature regarding community discovery is reviewed in Sect. 3.2. The fuzzy graph model and its fundamental properties are outlined along with two remarks about higher order data and partitioning algorithm termination criteria in Sect. 3.3. Fuzzy Walktrap and Fuzzy Newman-Girvan algorithms are presented in detail in Sects. 3.4 and 3.5 respectively. The Kronecker model is briefly reviewed in Sect. 3.8, where the results of the application of Fuzzy Walktrap and Fuzzy Newman-Girvan to a number of Kronecker synthetic graphs are reported. Finally, the main findings of this work as well as future research directions are discussed in Sect. 3.9. Table 3.1 summarizes the symbols used in this work. Notice that v_k and e_k are shorthand notations for the k-th vertex and the k-th edge respectively. The graph they refer to should be clear or implied by the context.

3.2 Related Work

Large graph community detection or community identification is significant with building big data analytics. This problem is algorithmically reduced to either graph partitioning or data clustering in general [4, 10, 22, 28]. Graph partitioning can be performed either structurally or spectrally. In the former case the graph is interpreted as an algebraic object and the partitioning is based on the properties of the graph adjacency matrix [19, 29], whereas in the latter the graph is treated as a combinatorial object and the partitioning exploits features such as edge density and community coherence [3, 22, 28].

A research problem related to community detection is authority estimation. In [1] several graph features, such as the degree distribution and hub and authority scores, are used for modelling the relative importance of a given user. Alternatively, in the expertise ranking model [12], authorities are derived by performing link analysis to the graph induced from interactions between users. More to the point, in [30] authors employ Latent Dirichlet Allocation and a PageRank variant to cluster the graph according to topics and in following the authorities for each topic are identified. This was extended in [24] with additional features, advanced clustering along with real-time applicability.

This view is echoed throughout previous works where several features of Twitter users are introduced so as to be taken into consideration by the community detection algorithms. In these cases, the proposed metrics can also explicitly describe the profile of a user regarding their popularity as well as the percentage of their perceived influence. Works regarding influential communities identification are presented in [13, 14], while works considering message diffusion within a given graph are [15, 16]. Moreover, [17, 18] deal with emotional modeling with respect to user influence [31].

In [15] the notion of influence from users to networks is extended and in following, personality as a key characteristic for identifying influential networks is considered. The system creates influential communities in a Twitter network graph by considering user personalities where an existing modularity-based community detection algorithm is used. At a later point, the insertion of a pre-processing step that eliminates graph edges based on user personality is utilized. In [17] a methodology for estimating the importance and the influence of a user in a Twitter network is described. Specifically, the authors propose a schema where users are represented by nodes and the edges, which connect these vertices, represent the relations of Follower to Following introduced by Twitter. Furthermore, in [31] some metrics for estimating user influence by presenting an analysis on the current strength of Twitter is proposed. Authors claim that the effect of influence is usually sighted when followers are affected via corresponding posts, even though the existence of this kind of friendship might be ignored.

Analytics are an integral part of large graph processing systems such as massive distributed graph computing systems like Google Pregel and graph based machine learning frameworks like Graphlab.[1] In these systems, graphs play a dual role as the computational flow model as well as the learning model. Interest in the graph processing field [23, 27] has been invigorated with the advent of open source graph databases such as BrightStar,[2] Neo4j,[3] Sparksee[4] and GraphDB.[5] Finally, fuzzy graphs have been introduced among others in [25], where a fuzzy extension of Cypher termed FUDGE is presented.

---

[1]https://dato.com/products/create/open_source.html.

[2]https://brightstardb.com.

[3]http://www.neo4j.com.

[4]http://www.sparsity-technologies.com.

[5]http://www.ontotext.com.

3.3 Fuzzy Graphs

3.3.1 Definitions

Within the context of this work, fuzzy graphs are combinatorial objects with a fixed set of vertices and a fuzzy set of edges. Its formal definition is the following.

Definition 3.1 A fuzzy graph is the ordered triplet

$$G = (V, E, h) \tag{3.1}$$

where $V = \{v_k\}$ is the set of vertices, $E = \{e_k\} \subseteq V \times V$ is the set of edges, and $h(\cdot)$ is the edge membership function

$$h : E \rightarrow (0, 1] \tag{3.2}$$

which measures the degree of participation of e_k to G. Notice that for brevity, non-existent edges are not included in the fuzzy graph.

It's obvious that a more concise definition of a fuzzy graph would need only V and h with the latter being zero for each vertex pair in the set $(V \times V) \setminus E$.

Definition 3.2 Under the fuzzy graph model the cost $\delta(e_k)$ of traversing e_k is

$$\delta(e_k) \triangleq \frac{1}{h(e_k)} \in [1, +\infty) \tag{3.3}$$

Besides (3.3) there are other relationships between $\delta(e_k)$ and $h(e_k)$, each of which is suitable for a particular subclass of problems. Common choices include

$$\delta(e_k) = \begin{cases} \eta_0, & 0 < h(e_k) \leq \tau_0 \\ \frac{1}{\sqrt{h(e_k)}}, & \tau_0 \leq h(e_k) \leq \tau_1 \\ \eta_1, & \tau_1 < h(e_k) \leq 1 \end{cases}$$

$$\delta(e_k) = \frac{1}{\gamma_0 + h^p(e_k)}, \quad \gamma_0 > 0, p \in \mathbb{Z}$$

$$\delta(e_k) = \prod_{j=1}^{n} \left(\frac{1}{\gamma_j + h^{p_j}(e_k)} \right)^{\frac{1}{n}}, \quad \gamma_j > 0, p_j \in \mathbb{Z}^+, 1 \leq j \leq n$$

$$\delta(e_k) = \prod_{j=1}^{n} \left(\frac{1}{\gamma_j + h^{p_j}(e_k)} \right)^{\frac{1}{\sum_{j=1}^{n} p_j}}, \quad \gamma_j > 0, p_j \in \mathbb{Z}^+, 1 \leq j \leq n$$

$$\delta(e_k) = \frac{1}{\alpha_0 + e^{-\frac{1}{\beta_0} h(e_k)}}, \quad \alpha_0, \beta_0 > 0$$

$$\delta(e_k) = \tanh\left(\frac{1}{\beta_0} h(e_k)\right), \quad \beta_0 > 0$$
$$\delta(e_k) = h(e_k) \log_b\left(\frac{1}{h(e_k)}\right), \quad b \geq 2$$
$$\delta(e_k) = \frac{1}{2}\left(h(e_k) + \frac{1}{h(e_k)}\right) \tag{3.4}$$

Consider that in each of the above cases $\delta_p(e_k)$ depends only on $h(e_k)$. This can be extended so that $\delta_p(e_k)$ may be defined as a function of edges, which are adjacent to either endpoint of e_k. However, this may lead to recursive computations for $\delta_p(e_k)$, making it essentially a global property instead of a local one. Particular choices of $\delta_p(e_k)$ might avoid global computations, but this topic is off the scope of this work.

Definition 3.3 The cost Δ_p of a path $p = (e_1, \ldots, e_m)$ of a fuzzy graph is the sum of the cost of its edges

$$\Delta_p \triangleq \sum_{e_k \in p} \delta(e_k) = \sum_{k=1}^{m} \frac{1}{h(e_k)} = \frac{1}{m\,\mathcal{H}\left(h(e_1), \ldots, h(e_m)\right)} \tag{3.5}$$

where $\mathcal{H}\left(h(e_1), \ldots, h(e_m)\right)$ denotes the harmonic mean of $h(e_k)$.

Observe that Δ_p is dominated by the minimum of $h(e_k)$, indicating that low cost paths contain exclusively edges that are highly likely to belong to the graph. Also loose upper and lower bounds for Δ_p are

$$\frac{1}{\max_{h(e_k)\in p}\left\{h(e_k)\right\}} \leq \Delta_p \leq \frac{1}{m} \cdot \frac{1}{\min_{h(e_k)\in p}\left\{h(e_k)\right\}} \tag{3.6}$$

Definition 3.4 The strength Σ_p of a path $p = (e_1, \ldots, e_m)$ of a fuzzy graph is defined as the minimum value the membership function takes in p

$$\Sigma_p \triangleq \min_{e_k \in p}\left\{h(e_k)\right\} \tag{3.7}$$

Definition 3.5 The distance $d(v_s, v_t)$ between v_s and v_t is defined as the minimum cost over all paths connecting them

$$d(v_s, v_t) \triangleq \min_{p}\left\{\Delta_p\right\}, \quad p = (v_s, v_1, \ldots, v_n, v_t) \tag{3.8}$$

In [25] it is established that $d(\cdot, \cdot)$ is a distance function, namely it satisfies the following properties

- $d(v_i, v_j) = 0 \Leftrightarrow v_j = v_i, \quad \forall v_i, v_j \in V$
- $d(v_i, v_j) = d(v_j, v_i), \quad \forall v_i, v_j \in V$
- $d(v_i, v_j) \leq d(v_i, v_k) + d(v_k, v_j), \quad \forall v_i, v_j, v_k \in V$

3.3.2 Weight Distributions

Although $h(\cdot)$ can be arbitrary, the fuzzy graphs which have been used in the experiments were chosen such as that $h(e_k)$ would be distributed according to the non-central χ_2^2 distribution with two degrees of freedom. Specifically, if g_1 and g_2 are identical and independently distributed so that

$$f(g_1) = \frac{1}{\sigma_0\sqrt{2\pi}} e^{-\frac{(g_1-\mu_0)^2}{2\sigma_0^2}} \tag{3.9}$$

and

$$f(g_2) = \frac{1}{\sigma_0\sqrt{2\pi}} e^{-\frac{(g_2-\mu_0)^2}{2\sigma_0^2}} \tag{3.10}$$

then the non-linear transformation

$$h_t(e_k) = \sqrt{\frac{g_1^2 + g_2^2}{2}}, \quad g_{1,2} \sim \mathcal{N}\left(\frac{1}{2}, \frac{1}{6}\right) \tag{3.11}$$

yields a noncentral χ_2^2 random variable. This distribution generates only positive values, a strict requirement in order to build a fuzzy graph, which are relatively concentrated around the mean value, less strongly than the Gaussian case though. The mean and the variance have been chosen as the interval

$$\left[\mu_0 - 3\sigma_0\,,\, \mu_0 + 3\sigma_0\right] \tag{3.12}$$

to coincide with $[0, 1]$.

Common choices for edge weight mass distributions include

- The binomial distribution (biological networks).
- The Poisson distribution (computer networks).
- The exponential distribution (computer networks).
- The geometric distribution (queue theory).
- The inverse geometric distribution (queue theory).
- The dgx distribution [2] (biological networks).

- The Zipf distribution (bibliographic networks).
- The multifractal distribution (brain connectomics).

while popular choices for edge weight discretized probability density function include

- The discretized lognormal distribution (signal processing).
- The discretized chi square distribution (signal processing).
- The discretized Weibull distribution (operations research).
- The discretized Rayleigh distribution (wireless communications).
- The discretized Rice distribution (wireless communications).

Take into consideration that a mass distribution function or a discretized probability density function must have strictly positive support in order to qualify as an edge weight distribution.

3.3.3 Elementary Quality Metrics of Fuzzy Graphs

Definitions 3.6 and 3.7 outline two important structural graph metrics, namely the *density* and the *completeness* of a fuzzy graph, used to quickly assess the structure of fuzzy graphs. Both indicate compression potential in different ways, as the former is defined as the ratio of edges to vertices, whereas the latter is the number of edges to the number of edges of K_n.

Definition 3.6 The (log)density ρ_0 (ρ_0') of a fuzzy graph $G(V, E, h)$ is the ratio of the (logarithm)number of its edges to the (logarithm)number of its vertices.

$$\rho_0 \stackrel{\triangle}{=} \frac{|E|}{|V|} \quad \text{and} \quad \rho_0' \stackrel{\triangle}{=} \frac{\log |E|}{\log |V|} \tag{3.13}$$

Definition 3.7 The (log)completeness σ_0 (σ_0') of a fuzzy graph $G(V, E, h)$ is the ratio of the (logarithm)number of its edges to the (logarithm)number of the edges of the complete graph with the same number of vertices.

$$\sigma_0 \stackrel{\triangle}{=} \frac{|E|}{\binom{|V|}{2}} \quad \text{and} \quad \sigma_0' \stackrel{\triangle}{=} \frac{\log |E|}{\log \binom{|V|}{2}} \approx \frac{\log |E|}{2\log |V|} = \frac{\rho_0'}{2} \tag{3.14}$$

3.3.4 Higher Order Data

The class of fuzzy graphs defined in this section is a typical example of higher order big data. In the community identification case the order of the data is expressed either in terms of the number of vertices involved in each community or in terms of the number of edges that need to be traversed in order to form a community.

An indication of the connection between community detection and the higher order analytics is that the smallest community is a triangle, either a closed or an open one. Observe that the closed triangle in terms of both vertices and edges is clearly a third order metric, whereas an open triangle is a third order metric only in terms of vertices. The link between community structure and higher order structure stems from the fact that isolated edges do not qualify as communities, since information exchange is trivial among its endpoints. In other words, in a given vertex group in order to qualify as a community there has to be at least one vertex connecting the remaining ones. This allows efficient information dissemination within the structure of a graph, usually in a way which balances local and global communication.

3.4 Fuzzy Walktrap

Community discovery algorithms in general have as input a graph and, possibly, a termination criterion. Although the graphs in this paper are fuzzy, the communities are deterministic in the sense that each vertex belongs only to one community. Hence, the vertex set V is partitioned as follows:

Definition 3.8 A (deterministic) partition of a set $S = \{s_1, \ldots, s_n\}$ is an assignment of elements s_k to p sets S_j such that $\cup_{j=1}^{p} S_k = S$ and $S_i \cap S_j = \emptyset$ for each distinct index pair i and j. Therefore

$$\sum_{j=1}^{p} |\, S_j \,| = |\, S \,| \tag{3.15}$$

A partition is denoted as $S = \{S_j\}$.

More advanced algorithms would partition V in a fuzzy sense as follows:

Definition 3.9 A fuzzy partition of a set $S = \{s_1, \ldots, s_n\}$ is an assignment of elements s_k to p sets S_j where the fuzzy membership operator $\in_F$ quantifies the degree of participation of s_k to S_j.

$$s_k \in_F S_j = \mu_{k,j} \in [0, 1] \tag{3.16}$$

A fuzzy partition is denoted as $S = \{S_j\}_F$. Notice that

$$\sum_{k=1}^{n} \sum_{j=1}^{p} \mu_{k,j} = |\, S \,| \tag{3.17}$$

It is perhaps important at this point to emphasize that, contrary to deterministic set partition, fuzzy partitioning creates subsets which are pairwise overlapping. The actual overlap magnitude depends on the number of subsets as well as on the selected membership function.

The Walktrap algorithm is based on the concept of an edge traversing random walker and is presented in Algorithm 1. The crucial observation is that although the walker starts from an arbitrary vertex, it will eventually spend more time in densely interconnected graph segments. This is based on the fact that it is more probable for a randomly picked edge to lead to another vertex inside the community the walker is currently in than to a vertex of another community [26]. The probability that the walker moves from v_i to v_j is

$$p_{i,j} = \frac{\mathbf{A}[i,j]}{\deg(v_i)} \tag{3.18}$$

where **A** denotes the adjacency matrix of the graph where

$$\mathbf{A}[i,j] \triangleq \begin{cases} 1, v_i = v_j \vee (v_i, v_j) \in E \\ 0, v_i \neq v_j \wedge (v_i, v_j) \notin E \end{cases} \in \{0,1\}^{|V|\times|V|} \tag{3.19}$$

If the probability that the random walker reaches v_j from v_i through a path of length ℓ is denoted by $p^{\ell}_{i,j}$, then the following should hold:

- If v_i and v_j belong to the same community, then $p^{\ell}_{i,j}$ should be large for at least large values of ℓ. Note that the converse is not always true. In other words, depending on graph topology $p^{\ell}_{i,j}$ may be high even if v_i and v_j belong to different communities.
- If v_i and v_j belong to the same community, then the random walker tends to treat them in a very similar manner. Then for each ℓ the condition $p^{\ell}_{i,j} = p^{\ell}_{j,i}$ should hold.
- It should be also noted that $p^{\ell}_{i,j}$ depends on $\deg(v_j)$, as the walker is more probable to select a higher degree vertex.

The above conditions lay the groundwork for defining the distance $d_{i,j}$ between v_i and v_j as

$$d_{i,j} \triangleq \sqrt{\sum_{k=1}^{|V|} \frac{\left(p^{\ell}_{i,k} - p^{\ell}_{j,k}\right)^2}{\deg(v_k)}} \tag{3.20}$$

In a similar manner, the transition probability $p^{\ell}_{C,k}$ from any vertex belonging to a community C to v_k in ℓ steps is defined as

$$p^{\ell}_{C,k} = \frac{1}{|C|} \sum_{i \in C} p^{\ell}_{i,k} \tag{3.21}$$

Generalizing (3.20), the distance r_{C_i,C_j} between the communities C_i and C_j is defined as

$$\rho_{i,j} = \sqrt{\sum_{k=1}^{|V|} \frac{\left(p^{\ell}_{C_i,k} - p^{\ell}_{C_j,k}\right)^2}{\deg\left(v_k\right)}} \tag{3.22}$$

Algorithm 1 Deterministic Walktrap

Require: Graph $G(V, E)$ and termination criterion γ_0
Ensure: G is partitioned into communities; $V = \{V_k\}$ is the community set
for all $v_k \in V$ **do**
  $V_k \leftarrow \{v_k\}$
end for
repeat
  for all distinct pairs $\left(V_i, V_j\right)$ **do**
    $\rho_{i,j} \leftarrow r_{V_i,V_j}$ **and** $\rho_{j,i} \leftarrow \rho_{i,j}$
  end for
  $(i^*, j^*) \leftarrow \mathrm{argmin}_{i,j}\{\rho_{i,j}\}$
  $V \leftarrow V \setminus V_{i*}$ **and** $V \leftarrow V \setminus V_{j*}$
  $p \leftarrow |V|$ **and** $V_{p+1} \leftarrow V_{i^*} \cup V_{j^*}$ **and** $V \leftarrow V \cup V_{p+1}$
until V satisfies γ_0
return V

The Fuzzy Walktrap begins by assigning each vertex to its own community like in the deterministic case and is presented in Algorithm 2. Notice that in the fuzzy case the entries of $\mathbf{A}^F$ now are

$$\mathbf{A}^F\left[i,j\right] \triangleq \begin{cases} 1, & v_i = v_j \\ h(\left(v_i, v_j\right)), & \left(v_i, v_j\right) \in E \\ 0, & \left(v_i, v_j\right) \notin E \end{cases} \in [0, 1]^{|V|\times|V|} \tag{3.23}$$

Therefore, they are continuous and belong to [0, 1] instead of being binary. The algorithm then proceeds as the deterministic Walktrap in order to locate communities with the distance metrics unaltered.

3.5 Fuzzy Newman-Girvan

The Newman-Girvan (Algorithm 3) or edge betweeness algorithm [11] is based on betweeness centrality, an edge centrality metric which counts the fraction of the number of the shortest paths connecting two vertices v_i and v_j a given edge e_k is part of, denoted by $\theta^k_{i,j}$, to the total number of shortest paths connecting v_i and v_j, denoted by $\theta_{i,j}$. Then the betweeness centrality for e_k, denoted by B_k, is computed by averaging over each vertex pair

Algorithm 2 Fuzzy Walktrap

Require: Fuzzy graph $G(V, E, h)$ and termination criterion γ_0
Ensure: G is partitioned into communities; $V = \{V_k\}$ is the community set

for all $v_k \in V$ **do**
  $V_k \leftarrow \{v_k\}$
end for
repeat
  for all distinct pairs (V_i, V_j) **do**
    $\rho^F_{i,j} \leftarrow r^F_{V_i,V_j}$ **and** $\rho^F_{j,i} \leftarrow \rho^F_{i,j}$
  end for
  $(i^*, j^*) \leftarrow \text{argmin}_{i,j}\left\{\rho^F_{i,j}\right\}$
  $V \leftarrow V \setminus V_{i*}$ **and** $V \leftarrow V \setminus V_{j*}$
  $p \leftarrow |V|$ **and** $V_{p+1} \leftarrow V_{i^*} \cup V_{j^*}$ **and** $V \leftarrow V \cup V_{p+1}$
until G satisfies γ_0
return V

$$B_k \triangleq \frac{1}{\binom{|V|}{2}} \sum_{(v_i,v_j)\in E} \left(\frac{\theta^k_{i,j}}{\theta_{i,j}}\right), \quad v_i \neq v_j \tag{3.24}$$

In [11] a process for computing B_k for each e_k in a manner resembling breadth-first search is described. The rationale is that vertices belonging to different communities should rely on edges connecting communities for information exchange. However, regard that the converse need not be true. Moreover, depending on graph topology, some of the community connecting edges may not rank high in terms betweeness centrality as other edges may be more preferable. Therefore, the edge e^* with the highest betweeness centrality should be removed and the process should be applied again to the new graph. Eventually, all edges connecting communities will be identified. Intuitively, the edge sequence $\langle e^* \rangle$ should contain the graph bridges as well, which are a subset of the community connecting edges. In case the graph becomes disconnected, then the process is repeated to each of the connected components.

Algorithm 3 Deterministic Newman-Girvan

Require: Graph $G(V, E)$ and termination criterion γ_0
Ensure: G is partitioned into communities; $V = \{V_k\}$

while $E \neq \varnothing$ **and** γ_0 **not** satisfied **do**
  compute shortest paths in G
  for all $e_k \in E$ **do**
    compute B_k as in (3.24)
  end for
  $e^* \leftarrow \text{argmax}_k\{B_k\}$
  $E \leftarrow E \setminus \{e^*\}$
end while
return V

The structure of Fuzzy Newman-Girvan (Algorithm 4) is identical to that of its deterministic counterpart. This happens because the fuzzy edge costs affect only the computation of B_k^F, the fuzzy edge betweeness centrality, and not the edge selection. It should be noted that for a computational point of view, any efficient algorithm for weighted graphs can be employed. Typically BFS is employed, as its average complexity is $\mathrm{O}(\mid V \mid + \mid E \mid)$, which reduces to $\mathrm{O}(\mid V \mid)$ for sparse graphs or to $\mathrm{O}\left(\mid V \mid^s\right)$ when $\mid E \mid = \mathrm{O}\left(\mid V \mid^s\right)$.

Algorithm 4 Fuzzy Newman-Girvan

Require: Fuzzy graph $G(V, E, h)$ and termination criterion γ_0
Ensure: G is partitioned into communities; $V = \{V_k\}_F$

```
while E ≠ ∅ and γ_0 not satisfied do
  compute shortest fuzzy paths in G
  for all e_k ∈ E do
    compute B_k^F as in (3.24)
  end for
  e* ← argmax_k {B_k^F}
  E ← E \ {e*}
end while
return V
```

3.6 Termination Criteria and Clustering Evaluation

Fuzzy Walktrap eventually terminates as in each step communities are merged, until the graph becomes one single community. Likewise, Fuzzy Newman-Girvan terminates when the communities degenerate to single vertices. Clearly, neither case is very useful and thus, any admissible termination criterion γ should yield at least two communities.

Termination criteria are not trivially selected. The most easy is to require the formation of a fixed number of communities k_0. However, as this cannot be known in advance, it is usually selected empirically. Common choices include

$$\log \mid V \mid \quad \text{and} \quad \sqrt[q]{\mid V \mid} \quad \text{and} \quad \sqrt[q]{\frac{\mid V \mid}{\mid E \mid}} \tag{3.25}$$

The ratio of average community diameter to the graph diameter is presented below. If $\left\{C_k\right\}_{k=1}^p$ is the community set, ζ_k the diameter of C_k, and ζ_0 the graph diameter then

$$\gamma_1 \stackrel{\triangle}{=} \frac{1}{p\zeta_0} \sum_{k=1}^{p} \zeta_k \neq \sum_{k=1}^{p} \frac{\zeta_k}{\zeta_0} \tag{3.26}$$

When γ_1 becomes a fraction of ζ_0, then the algorithm terminates.

Let V_k and E_k denote the vertices and edges inside C_k. Then, another termination criterion is the average ratio of $\mid E_k \mid$ to the number of edges of the fully connected graph of the same number of vertices $\mathrm{K}_{|V_k|}$

$$\gamma_2 \triangleq \frac{1}{p}\sum_{k=1}^{p} \frac{\mid E_k \mid}{\binom{|V_k|}{2}} = \frac{2}{p}\sum_{k=1}^{p} \frac{\mid E_k \mid}{\mid V_k \mid (\mid V_k \mid - 1)} \approx \frac{2}{p}\sum_{k=1}^{p} \frac{\mid E_k \mid}{\mid V_k \mid^2} \tag{3.27}$$

The algorithm terminates when γ_2 becomes a fraction of global graph density.

Alternatively, the Tanimoto similarity measure can be used to assess the pairwise similarity of the clusters generated in two consecutive clustering iterations. For two sets S_1 and S_2 the Tanimoto coefficient τ_{S_1,S_2} is defined as

$$\tau_{S_1,S_2} \triangleq \frac{\mid S_1 \cap S_2 \mid}{\mid S_1 \cup S_2 \mid} = \frac{\mid S_1 \cap S_2 \mid}{\mid S_1 \mid + \mid S_2 \mid - \mid S_1 \cap S_2 \mid}, \quad 0 \leq \tau_{S_1,S_2} \leq 1 \tag{3.28}$$

If $S^{[j]} \triangleq \left\{ C_k^{[j]} \right\}$ is the community set generated during iteration j, then

$$\gamma_3 \triangleq \frac{1}{\mid S^{[j]} \mid\mid S^{[j-1]} \mid} \sum_{k=1}^{|S^{[j]}|} \sum_{k'=1}^{|S^{[j-1]}|} \tau_{C_k^{[j]}, C_{k'}^{[j-1]}} \tag{3.29}$$

is a termination criterion. Similarly to γ_1 and γ_2, γ_3 also requires a threshold.

Observe that γ_3 can potentially outperform γ_1 and γ_2, as it relies on two clustering iterations. In order to incorporate a similar logic to γ_1 and γ_2 as well as to any termination criterion γ based on knowledge from a single iteration, the change rate between successive clustering iterations can be used. If $\left\langle \gamma^{[j]} \right\rangle$ is the sequence of the values of γ in each iteration, then

$$\eta_0^{[j+1]} \triangleq \mid \gamma^{[j+1]} - \gamma^{[j]} \mid \tag{3.30}$$

should be kept bounded. Similarly, $\left\langle \gamma^{[j]} \right\rangle$ change over two successive clustering iterations can be measured as

$$\eta_1^{[j+1]} \triangleq \frac{\mid \gamma^{[j+1]} - 2\gamma^{[j]} + \gamma^{[j-1]} \mid}{2} \leq \frac{1}{2}\left(\eta_0^{[j+1]} + \eta_0^{[j]} \right) \tag{3.31}$$

Once clustering algorithm is executed, its results should be evaluated. The scoring system used in this article is the ratio of magnitude order of the clustering score q_0 to the magnitude order of the total execution time q_0.

$$w_0 \triangleq \frac{\log\left(1 + \frac{1}{1+q_0}\right)}{\log\left(1 + t_0\right)} = \left(\sum_{k=1}^{+\infty} \frac{1}{k\left(1+q_0\right)^k} \right) \left(\sum_{k=1}^{+\infty} \frac{t_0^k}{k} \right)^{-1} \approx \frac{1}{t_0\left(1+q_0\right)} \tag{3.32}$$

Thus, w_0 maintains a balance between clustering quality and clustering cost. The approximation formula results from keeping the first order Taylor term of both nominator and denominator and it was not used in Sect. 3.8.

Clustering quality was assessed by the fuzzy average edge cut, defined as follows. By fixing a community C^* from the community set S, the graph is partitioned to two subgraphs G_1 and G_2. The fuzzy edge cut cost for a particular community C^* is the ratio of the sum of the fuzzy cost of the edges whose end vertices belong to different subgraphs to the sum of the fuzzy cost of the edges whose end vertices belong to the same subgraph. Averaging over each community $C^* \in S$

$$q_0 \triangleq \frac{1}{|S|} \sum_{C^* \in S} \frac{\sum_{p \in G_1 \wedge q \in G_2} \delta((p,q))}{\sum_{p,q \in G_1} \delta((p,q)) + \sum_{p,q \in G_2} \delta((p,q))} \tag{3.33}$$

Since a low value of q_0 implies a high quality graph partitioning, the nominator of (3.32) is inversely proportional to q_0.

3.7 Source Code

This section provides a brief overview of the source code, placing emphasis either on the Neo4j API or on the programming techniques used to represent data retrieved from or inserted to a graph database and facilitate their processing.

The following Java libraries provide methods for parsing the input files:

```
import java.io.File;
import java.util.Scanner;
```

Classes, methods, and data types for interfacing with Neo4j are imported with the following libraries:

```
import org.neo4j.graphdb.DynamicLabel;
import org.neo4j.graphdb.GraphDatabaseService;
import org.neo4j.graphdb.Label;
import org.neo4j.graphdb.Node;
import org.neo4j.graphdb.Relationship;
import org.neo4j.graphdb.RelationshipType;
import org.neo4j.graphdb.Transaction;
import org.neo4j.graphdb.factory.GraphDatabaseFactory;
```

The single Neo4j instance was run in embedded mode, where the server and the client operate not only on the same machine but also on the same JVM. This provides direct access to the database at the expense of extensive database locks. For the definition of the EmbeddedNeo4j class, the following data types are of interest:

```
public class EmbeddedNeo4j {

  //...
  GraphDatabaseService GraphDB;
  Node cNode[] = new Node[arraySize];
  Relationship relationship;

  //...
  private static enum RelTypes implements RelationshipType
  {
    KNOWS
  }

  //...
  private static String[] EdgeProperty =
    new String[]{
      "probability" ,
      "cost"
    };
  private static final int ProbIndex = 0;
  private static final int CostIndex = 1;

  //...
}
```

The GraphDataBaseService is the actual database process that needs to be started, queried, and properly shut down. Data types Node and Relationship are used to build a graph as their names suggest. Moreover, the abstract enumeration type RelationshipType has to be implemented with the specific relationships of the particular graph. A token relationship KNOWS has been implemented. Edge probabilities and costs are stored as separate properties, allowing the cost computation under different fuzzy models.

The Main method is remarkably simple as it creates and populates the database, runs the fuzzy community detection algorithms, and terminates the database.

```
public static void main(final String[] args) {
  EmbeddedNeo4j neo4j = new EmbeddedNeo4j();
  neo4j.createDb();
  neo4j.createRelations();
  neo4j.fuzzyWalktrap();
  neo4j.fuzzyNewmanGirvan();
  neo4j.shutDown();
}
```

Method createDb populates the database in a single transaction. Since the graph is relatively small in size and no other processes are using the database, the lock time is small and has no consequence to other computations. A simple iteration in a **while** loop reads all vertices and from a .csv file and progressively inserts them to the database.

```
void createDb() {

  //...
  graphDb = new GraphDatabaseFactory();
    newEmbeddedDatabase(DBPath);
  registerShutdownHook(graphDb);

  //...
  try (Transaction tx = graphDb.beginTx()) {
    Label NodeLabel = DynamicLabel.label(DBLabel);
    try {
      Scanner x = new Scanner(metrics);
      while (x.hasNext()) {
        //...
        cNode[j] = graphDb.createNode();
        cNode[j].addLabel(NodeLabel);
      }
      x.close();
    }

    //...
    catch(Exception e) {
      System.out.println(metrics.getAbsolutePath());
    }

    //...
    tx.success();
  }
}
```

In a similar manner, the createRelations method parses the adjacency table of the same .csv file in a single transaction and sets the appropriate values for edge properties.

```
void createRelations() {

  //...
  try {
    Transaction tx = graphDb.beginTx();
    try {
      Scanner x = new Scanner(table);
      while(x.hasNext()) {
        text = x.next();
        if (text.equals(RelationshipFile)) {
```

```
                relationship =
                  cNode[j].createRelationshipTo
                  (cNode[i], RelTypes.KNOWS);
                cEdge[k].setProperty(EdgeProperty[ProbIndex], x.next());
                cEdge[k].setProperty(EdgeProperty[CostIndex], x.next());
              }
            }
          }
        }
        tx.succsess();

        //...
        catch (Exception e) {
          System.out.println(table.getAbsolutePath());
        }
      }
```

Finally, the shutDown method closes the database using the shutdown method provided by the Neo4j API.

```
void shutDown() {
    graphDb.shutdown();
}
```

3.8 Results

3.8.1 Data Summary

In order to evaluate experimentally the algorithms of Sects. 3.4 and 3.5, two graphs have been used, the synthetic Kronecker graph used in [8] and a much larger one, which is a segment of the SNAP Higgs graph.

The Kronecker synthetic graph generation model [21] relies on the adjacency matrix of a graph and on the recursive nature of scale free graphs in order to progressively build a large graph from smaller similar ones. Given a preselected original adjacency matrix **A**, the following graph sequence is generated

$$\begin{aligned} \mathbf{A}_0 &= \mathbf{A} \\ \mathbf{A}_{k+1} &= \mathbf{A}_k \otimes \mathbf{A}, \quad k \geq 1 \end{aligned} \tag{3.34}$$

The generator matrix $\mathbf{A}_0$ for the specific implementation is

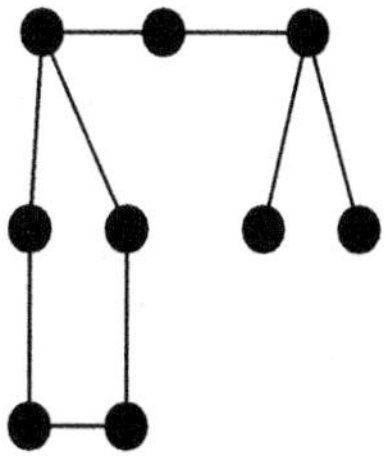

Fig. 3.1 Generator graph $\mathbf{A}_0$

Table 3.2 Synthetic graph $\mathbf{A}_4$ summary

Property	Value	Property	Value
Vertices	6561	Sample mean	0.3311
Edges	6561	Sample variance	0.1322
Density	1	Minimum probability	0.1332
Completeness	4.64×10^{-8}	Maximum probability	0.9701

$$\mathbf{A}_0 = \left[\begin{array}{ccccc|c|ccc} 1&1&0&0&1&1&0&0&0\\ 1&1&1&0&0&0&0&0&0\\ 0&1&1&1&0&0&0&0&0\\ 0&0&1&1&1&0&0&0&0\\ 1&0&0&1&1&0&0&0&0\\ \hline 1&0&0&0&0&1&0&0&1\\ \hline 0&0&0&0&0&0&1&0&1\\ 0&0&0&0&0&0&0&1&1\\ 0&0&0&0&0&1&1&1&1 \end{array}\right] = \left[\begin{array}{c|c|c} \mathbf{C}_1 & \mathbf{P}_{1,2} & \mathbf{P}_{1,3}\\ \hline \mathbf{P}^T_{1,2} & \mathbf{C}_2 & \mathbf{P}_{2,3}\\ \hline \mathbf{P}^T_{1,3} & \mathbf{P}^T_{2,3} & \mathbf{C}_3 \end{array}\right] = \left[\begin{array}{c|c|c} \mathbf{C}_1 & \mathbf{P}_{1,2} & \mathbf{O}\\ \hline \mathbf{P}^T_{1,2} & \mathbf{C}_2 & \mathbf{P}_{2,3}\\ \hline \mathbf{O} & \mathbf{P}^T_{2,3} & \mathbf{C}_3 \end{array}\right] \tag{3.35}$$

which contains two communities, namely $\mathbf{C}_1$ and $\mathbf{C}_3$, and the bridge $\mathbf{C}_2$ consisting of a single vertex which connects $\mathbf{C}_1$ and $\mathbf{C}_3$. $\mathbf{P}_{1,2}$, $\mathbf{P}_{1,3}$, and $\mathbf{P}_{2,3}$ contain the connection edges between these communities. Notice that $\mathbf{C}_1$ is a circle while $\mathbf{C}_3$ is a binary tree with a root and two vertices. Since $\mathbf{C}_1$ connects to $\mathbf{C}_3$ through $\mathbf{C}_2$, $\mathbf{P}_{1,3}$ is a zero block. $\mathbf{A}_0$ is shown in Fig. 3.1.

The Kronecker model was executed in four steps. Once the graph has been created, h_t is applied to its edges in order to make it fuzzy. Note that both $\mathbf{A}_0$ and $\mathbf{A}_4$ are moderately sparse. Moreover, the edge probabilities are closely concentrated around the mean value. Therefore, both from a structural and from an edge weight point of view, only a small number of communities is expected to be discovered by any well performing algorithm. Table 3.2 presents basic properties of $\mathbf{A}_4$. The left column has structural characteristics, while the right column refers to the actual probability values.

The SNAP Higgs graph [5] has been assembled from tweets, retweets, metions, and replies right before and immediately after the official announcement of Higgs boson by CERN. In this paper, a subgraph of Higgs graph was generated by keep-

Table 3.3 Modified Higgs dataset summary

Property	Value	Property	Value
Vertices	26718	Sample mean	0.3334
Edges	83145	Sample variance	0.1293
Density	3.1119	Minimum probability	0.0512
Completeness	2.32×10^{-4}	Maximum probability	0.9891

Table 3.4 t_0, q_0, and w_0 for each criterion, algorithm, and dataset

Algorithm	Criterion	t_0^K	q_0^K	w_0^K	t_0^H	q_0^H	w_0^H
FW	γ_1	2.1031	2.3412	0.2312	62.2008	9.1209	0.0227
FW	$\gamma_1 + \eta_1$	2.2260	1.9720	0.2476	65.7120	8.8432	0.0230
FW	$\gamma_1 + \eta_2$	3.2270	2.0011	0.1995	72.1410	8.8624	0.0225
FW	γ_2	2.0019	1.6272	0.2934	62.2002	8.8213	0.0234
FW	$\gamma_2 + \eta_1$	2.1276	1.5483	0.2903	65.7124	8.5203	0.0238
FW	$\gamma_2 + \eta_2$	2.2072	1.6210	0.2773	72.1415	8.5200	0.0233
FW	γ_3	2.8211	2.2099	0.2023	62.2002	9.0991	0.0228
FN-G	γ_1	6.0211	2.5309	0.1280	261.0034	7.1454	0.0208
FN-G	$\gamma_1 + \eta_1$	6.7222	2.8721	0.1124	258.8863	7.9813	0.0190
FN-G	$\gamma_1 + \eta_2$	9.0157	2.8721	0.0997	257.9912	7.9619	0.0190
FN-G	γ_2	5.9012	1.8281	0.1567	285.2203	6.7910	0.0213
FN-G	$\gamma_2 + \eta_1$	6.5823	1.8102	0.1503	281.4112	6.7734	0.0214
FN-G	$\gamma_2 + \eta_2$	8.9963	1.8101	0.1322	282.2532	6.7805	0.0214
FN-G	γ_3	14.3126	2.8889	0.0839	311.2302	6.7823	0.0210

ing only the *follow* relationships between Twitter accounts and converting them to undirected relationships, resulting thus in the modified Higgs dataset. Then, h_t was sampled as before. The properties of this dataset are shown in Table 3.3.

3.8.2 Analysis

Table 3.4 shows each algorithm and each dataset the time t_0, the average cut fuzzy cost q_0, and clustering score w_0 first for the Kronecker synthetic graph and then for the Higgs subgraph. Fuzzy Newman-Girvan is slower than Fuzzy Walktrap in all cases with a ratio which appears to decrease and graph size increases. Also, γ_1, the diameter ratio algorithm termination criterion, caused a slowdown. Although more experiments could shed light into this phenomenon, it should probably be attributed to the global nature of diameter computation. Edge density on the contrary is based exclusively on local properties and, thus, local data stored in cache are efficiently utilized. Furthermore, even a community diameter is algorithmically more expensive

Table 3.5 Communities for each criterion, algorithm, and dataset

Kronecker	γ_1	$\gamma_1+\eta_1$	$\gamma_1+\eta_2$	γ_2	$\gamma_2+\eta_1$	$\gamma_2+\eta_2$	γ_3
FW	22	21	21	23	23	22	24
FN-G	20	19	19	22	21	21	19
Higgs	γ_1	$\gamma_1+\eta_1$	$\gamma_1+\eta_2$	γ_2	$\gamma_2+\eta_1$	$\gamma_2+\eta_2$	γ_3
FW	1618	1601	1609	1633	1625	1618	1599
FN-G	1534	1526	1519	1542	1527	1523	1525

to compute compared to edge counting. Concerning the use of η_1 and η_2 as secondary termination criteria, they increase computational time and, although they improve the average cut cost, their score is lower.

The number of communities is shown in Table 3.5. In general, all criteria tend to give similar results for each graph. As a rule, γ_1 results in fewer communities. This can be explained as it is a global metric and, hence, communities are progressively constructed according to a common rule. On the contrary, γ_2 results in more communities being a local metric. Still, given the values of Table 3.4, the overall quality is not degraded. Finally, for the large Higgs dataset γ_3 is the most conservative, while for the Kronecker graph γ_3 generates more communities. A possible interpretation can be found in the richer connectivity patterns of the Higgs dataset or in the different scale of these graphs. Both factors result during early iterations in communities with high variability. The latter is gradually smoothed, leading to less communities at the expense of additional time.

Concerning the secondary criteria, there is a tendency for both η_1 and η_2 to generate fewer communities. Thus, the overhead shown in Table 3.4 should be attributed to extra computations and memory requirements.

In Figs. 3.2 and 3.3 are depicted the normalized sizes of the largest 16 communities detected by Fuzzy Walktrap and Fuzzy Newman-Girvan in the Kronecker and the Higgs datasets respectively. This information is repeated in Tables 3.6 and 3.7 for clarity. Numbering depends only on size and, thus, the fact that two communities are on the same row does not imply equivalence. The following analysis refers to the normalized community sizes.

Both algorithms in each dataset exhibit similar behavior with slight variations. They generate very clustered community sizes in the Higgs dataset to the point that the last ten communities have almost the same normalized size. Regarding the Kronecker dataset, Fig. 3.2 has also clustered community sizes of greater variability which seem to follow a Zipf distribution. To test deviation from the Zipf distribution, the logarithmic mean square error is used [21]

$$\xi = \left(\frac{1}{p} \sum_{k=1}^{p} \left(\log |C_k| - \left(\log \hat{\alpha}_{\mathrm{LS}} - \log k \hat{\gamma}_{\mathrm{LS}} \right) \right)^2 \right)^{\frac{1}{2}} \tag{3.36}$$

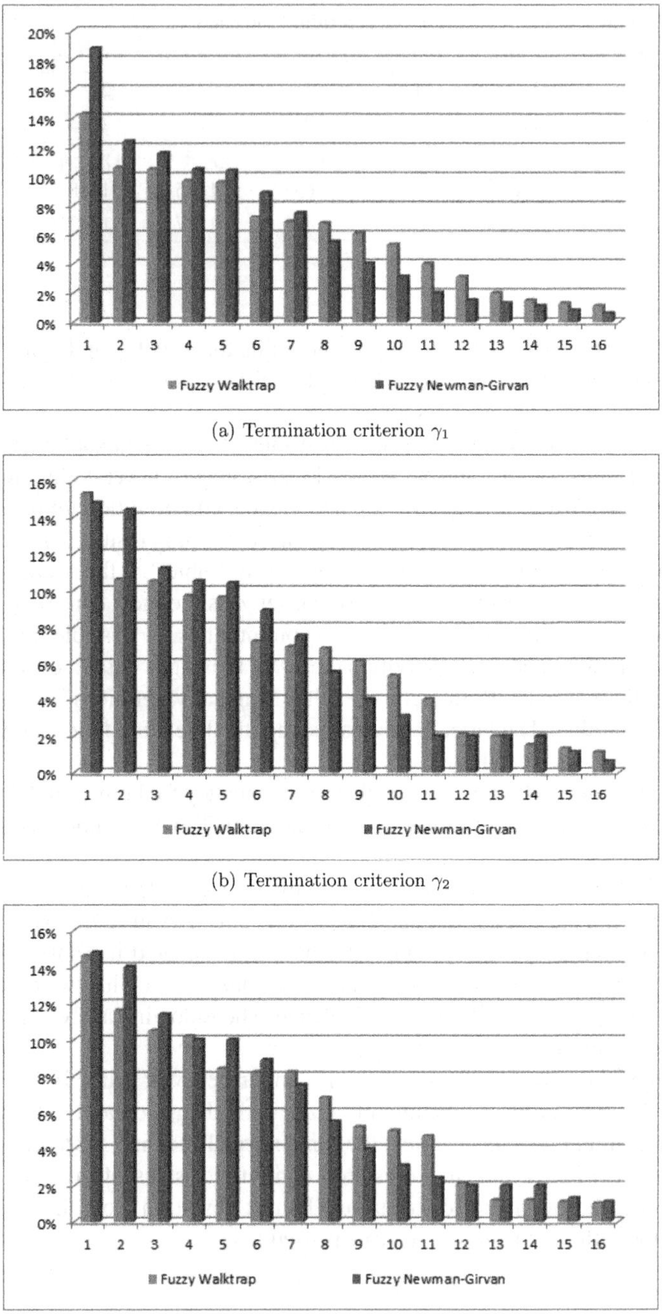

(a) Termination criterion γ_1

(b) Termination criterion γ_2

(c) Termination criterion γ_3

Fig. 3.2 Normalized community sizes for the Kronecker dataset

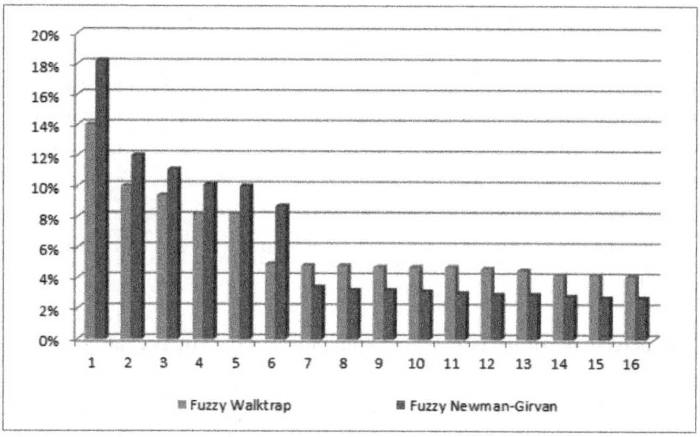

(a) Termination criterion γ_1

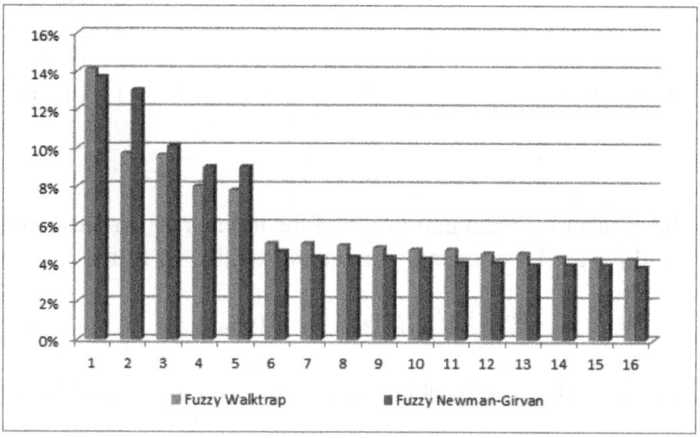

(b) Termination criterion γ_2

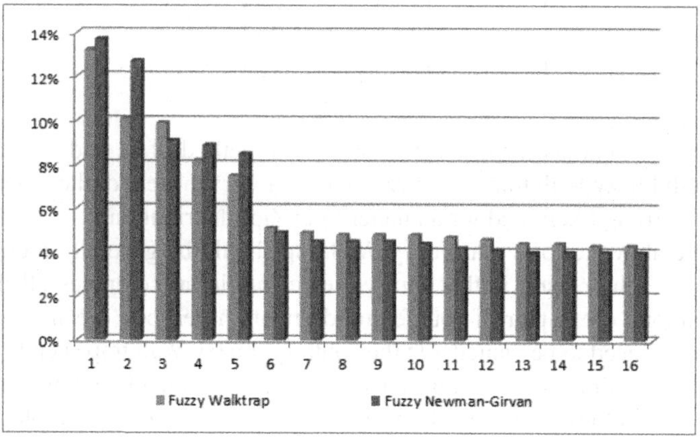

(c) Termination criterion γ_3

Fig. 3.3 Normalized community sizes for the Higgs dataset

Table 3.6 Size percentage of the largest 16 communities of the Kronecker dataset

Alg	Cr																
FW	γ_1	14.3	10.6	10.5	9.7	9.6	7.2	6.9	6.8	6.1	5.3	4.0	3.1	2.0	1.5	1.3	1.1
FW	γ_2	15.3	10.6	10.5	9.7	9.6	7.2	6.9	6.8	6.1	5.3	4.0	2.1	2.0	1.5	1.3	1.1
FW	γ_3	14.6	11.6	10.5	10.2	8.4	8.2	8.2	6.8	5.2	5.0	4.7	2.1	1.2	1.2	1.1	1.0
FN-G	γ_1	18.8	12.4	11.6	10.5	10.4	8.9	7.5	5.5	4.0	3.1	2.0	1.5	1.3	1.1	0.8	0.6
FN-G	γ_2	14.8	14.4	11.2	10.5	10.4	8.9	7.5	5.5	4.0	3.1	2.0	2.0	2.0	2.0	1.1	0.6
FN-G	γ_3	14.8	14.0	11.4	10.0	10.0	8.9	7.5	5.5	4.0	3.1	2.4	2.0	2.0	2.0	1.3	1.1

Table 3.7 Size percentage of the largest 16 communities of the Higgs dataset

Alg	Cr																
FW	γ_1	14.0	10.0	9.4	8.2	8.2	4.9	4.8	4.8	4.7	4.7	4.7	4.6	4.5	4.2	4.2	4.1
FW	γ_2	14.1	9.7	9.6	8.0	7.8	5.0	5.0	4.9	4.8	4.7	4.7	4.5	4.5	4.3	4.2	4.2
FW	γ_3	13.2	10.1	9.9	8.2	7.5	5.1	4.9	4.8	4.8	4.8	4.7	4.6	4.4	4.4	4.3	4.3
FN-G	γ_1	18.2	12.0	11.1	10.1	10.0	8.7	3.4	3.2	3.2	3.1	3.0	2.9	2.9	2.8	2.7	2.7
FN-G	γ_2	13.7	13.0	10.1	9.0	9.0	4.6	4.3	4.3	4.3	4.2	4.0	4.0	3.9	3.9	3.9	3.8
FN-G	γ_3	13.7	12.7	9.1	8.9	8.5	4.9	4.5	4.5	4.5	4.4	4.2	4.1	4.0	4.0	4.0	4.0

where p is the number of communities, $\mid C_k \mid$ the number of vertices in community C_k, and $\hat{\alpha}_{\mathrm{LS}}$ and $\hat{\gamma}_{\mathrm{LS}}$ are least squares estimators resulting by fitting

$$\mid C_k \mid \; = \; \alpha_0 \, k^{-\gamma_0}, \qquad \alpha_0 > 0, \gamma_0 \geq 1 \tag{3.37}$$

to community sizes. They can be obtained from the overdetermined linear system

$$\begin{bmatrix} \log \mid C_1 \mid \\ \log \mid C_2 \mid \\ \vdots \\ \log \mid C_p \mid \end{bmatrix} = \begin{bmatrix} 1 | 0 \\ 1 | -\log 2 \\ \ldots | \ldots \\ 1 | -\log p \end{bmatrix} \begin{bmatrix} \log \alpha_0 \\ \gamma_0 \end{bmatrix} \tag{3.38}$$

Notice that (3.36) holds for any other estimation method of α_0 and γ_0. The values for ξ for both Fuzzy Walktrap and Fuzzy Newman-Girvan were of the order of 10^{-2}, indicating a strong likelihood for an underlying Zipf distribution.

From the above follows that Fuzzy Walktrap should be generally preferable for larger graphs. The reason is that Fuzzy Newman-Girvan computes all the shortest paths among every pair of source and destination vertices, meaning that large graph portions need to be visited. On the contrary, Fuzzy Walktrap spends a considerable amount of time crossing adjacent edges before eventually moving on to other communities, exploiting thus local information. Moreover, from an implementation viewpoint Fuzzy Walktrap is preferrable, as it relies on local graph segments. Thus, fewer graph segments need to be loaded to memory, whereas the systematic traversal required to find shortest paths translates to more read and synchronization operations. Therefore, Fuzzy Walktrap may be the algorithm of choice in distributed graph processing systems [7, 9, 20].

3.9 Conclusions and Future Work

The primary contribution of this work is Fuzzy Walktrap and Fuzzy Newman-Girvan, two fuzzy graph community discovery algorithms based on their deterministic namesakes. The underlying fuzzy graph model was the one used in [25], where vertices are fixed while edges exist according to a probability distribution. Experimental results conducted on a synthetic fuzzy Kronecker graph as well as on a subset of the SNAP Higgs graph indicate that Fuzzy Newman-Girvan is considerably more expensive and it returns fewer communitie, while Fuzzy Walktrap is quicker and it makes efficient use of local information.

In future directions the development of more sophisticated fuzzy community identification algorithms should be included. Such algorithms could possibly take advantage of the specific membership function of a given graph or at least of some partial knowledge regarding a particular membership function in order to build heuristics for computation acceleration. For instance, knowledge about the first non-central moment of the membership function would help such an algorithm decide whether the cost of an edge is unusually high. Notice that both Fuzzy Walktrap and Fuzzy Newman-Girvan are distribution oblivious, so in essence they treat a fuzzy graph more like an ordinary weighted graph.

Another direction is the extension of the fuzzy model such that vertices would also participate to the graph according to a vertex membership function. Therefore, edge participation would be affected not only by the edge membership function but also by the vertex membership values of the two adjacent vertices.

References

1. Agichtein, E., Castillo, C., Donato, D., Gionis, A., Mishne, D.: Finding high-quality content in social media. In: Web Search and Data Mining Conference, WSDM, pp. 183–194. ACM (2008)
2. Bi, Z., Faloutsos, C., Korn, F.: The "dgx" distribution for mining massive, skewed data. In: Proceedings of the Seventh ACM SIGKDD International Conference on Knowledge Discovery and Data Mining, KDD'01, pp. 17–26. ACM, New York, NY, USA (2001)
3. Blondel, V.D., Guillaume, J.-L., Lambiotte, R., Lefebvre, E.: Fast unfolding of community hierarchies in large networks. J. Stat. Mech.: Theory Exp. **P1000** (2008)
4. Carrington, P.J., Scott, J., Wasserman, S.: Models and Methods in Social Network Analysis. Cambridge University Press (2005)

5. De Domenico, M., Lima, A., Mougel, P., Musolesi, M.: The anatomy of a scientific rumor. Sci. Rep. **3** (2013)
6. Deitrick W, Valyou B, Jones W, Timian J, Hu W (2013) Enhancing sentiment analysis on twitter using community detection. Commun. Netw. 5(3):192–197
7. Drakopoulos, G., Baroutiadi, A., Megalooikonomou, V.: Higher order graph centrality measures for Neo4j. In: Proceedings of the 6th International Conference of Information, Intelligence, Systems, and Applications, IISA 2015. IEEE (2015)
8. Drakopoulos, G., Kanavos, A., Makris, C., Megalooikonomou, V.: On converting community detection algorithms for fuzzy graphs in Neo4j. In: Proceedings of the 5th International Workshop on Combinations of Intelligent Methods and Applications, CIMA 2015. IEEE (2015)
9. Drakopoulos, G., Megalooikonomou, V.: On the weight sparsity of multilayer perceptrons. In: Proceedings of the 6th International Conference of Information, Intelligence, Systems, and Applications, IISA 2015. IEEE (2015)
10. Fortunato S (2010) Community detection in graphs. Phys. Rep. 486:75–174
11. Girvan M, Newman M (2002) Community structure in social and biological networks. Proc. Natl. Acad. Sci. 99(2):7821–7826
12. Jurczyk, P., Agichtein, E.: Discovering authorities in question answer communities by using link analysis. In: Conference of Information and Knowledge Management, CIKM, pp. 919–922 (2007)
13. Kafeza, E., Kanavos, A., Makris, C., Chiu, D.: Identifying personality-based communities in social networks. In: Legal and Social Aspects in Web Modeling (Keynote Speech) in Conjunction with the International Conference on Conceptual Modeling (ER), LSAWM (2013)
14. Kafeza, E., Kanavos, A., Makris, C., Vikatos, P.: Predicting information diffusion patterns in twitter. In: Artificial Intelligence Applications and Innovations, AIAI, pp. 79–89 (2014)
15. Kafeza, E., Kanavos, A., Makris, C., Vikatos, P.: T-PICE: Twitter personality based influential communities extraction system. In: IEEE International Congress on Big Data, pp. 212–219 (2014)
16. Kanavos, A., Perikos, I.: Towards detecting emotional communities in twitter. In: 9th IEEE International Conference on Research Challenges in Information Science, RCIS, pp. 524–525 (2015)
17. Kanavos, A., Perikos, I., Vikatos, P., Hatzilygeroudis, I., Makris, C., Tsakalidis, A.: Conversation emotional modeling in social networks. In: 26th IEEE International Conference on Tools with Artificial Intelligence, ICTAI, pp. 478–484 (2014)
18. Kanavos, A., Perikos, I., Vikatos, P., Hatzilygeroudis, I., Makris, C., Tsakalidis, A.: Modeling retweet diffusion using emotional content. In: Artificial Intelligence Applications and Innovations, AIAI, pp. 101–110 (2014)
19. Kernighan B, Lin S (1970) An efficient heuristic procedure for partitioning graphs. Bell Syst. Tech. J. 49(1):291–307
20. Kontopoulos, S., Drakopoulos, G.: A space efficient scheme for graph representation. In: Proceedings of the 26th International Conference on Tools with Artificial Intelligence, ICTAI 2014, pp. 299–303. IEEE (2014)
21. Leskovec J, Chakrabarti D, Kleinberg J, Faloutsos C, Ghahramani Z (2010) Kronecker graphs: an approach to modeling networks. J. Mach. Learn. Res. 11:985–1042
22. Newman, M.: Networks: An Introduction. Oxford University Press (2010)
23. Panzarino, O.: Learning Cypher. PACKT Publishing (2014)
24. Pal, A., Counts, S.: Identifying topical authorities in microblogs. In: Web Search and Data Mining, WSDM, pp. 45–54 (2011)
25. Pivert, O., Thion, V., Jaudoin, H., Smits, G.: On a fuzzy algebra for querying graph databases. In: Proceedings of the 26th International Conference on Tools with Artificial Intelligence, ICTAI 2014, pp. 748–755. IEEE (2014)
26. Pons, P., Latapy, M.: Computing communities in large networks using random walks. arXiv:physics/0512106 (2005)
27. Robinson, I., Webber, J., Eifrem, E.: Graph Databases. O'Reilly (2013)
28. Scott, J.: Social Network Analysis: A Handbook. SAGE Publications Ltd. (2000)

29. Shi J, Malik J (2000) Normalized cuts and image segmentation. IEEE Trans. Pattern Anal. Mach. Intell. 22(8):888–905
30. Weng, J., Lim, E.-P., Lim, J., Jiang, Q.H.: Twitterrank: Finding topic-sensitive influential twitterers. In: Web Search and Data Mining, WSDM, pp. 261–270 (2010)
31. Zamparas, V., Kanavos, A., Makris, C.: Real time analytics for measuring user influence on twitter. In: 27th IEEE International Conference on Tools with Artificial Intelligence, ICTAI (2015)

Chapter 4
Difficulty Estimation of Exercises on Tree-Based Search Algorithms Using Neuro-Fuzzy and Neuro-Symbolic Approaches

Foteini Grivokostopoulou, Isidoros Perikos and Ioannis Hatzilygeroudis

Abstract A central topic in the educational process concerns the engagement and involvement of students in educational activities that are tailored and adapted to their knowledge level. In order to provide exercises and learning activities of appropriate difficulty, their difficulty level should be accurately and consistently determined. In this work, we present a neuro-fuzzy and a neuro-symbolic approaches that are used to determine the difficulty level of exercises on tree-based search algorithms and we examine their performance. For the estimation of the difficulty level of the exercises, parameters like the number of the nodes of the tree, the number of children of each node, the maximum depth of the tree and the length of the solution path are taken into account. An extensive evaluation study was conducted on a wide range of exercises for blind and heuristic search algorithms. The performance of the approaches has been examined and compared against that of expert tutors. The results indicate quite promising performance and show that both approaches are reliable, efficient and confirm the quality of their exercise difficulty identification.

Keywords Exercise difficulty estimation · Search algorithms · Tree analysis · Neuro-fuzzy approach · Neurules

F. Grivokostopoulou (✉) · I. Perikos · I. Hatzilygeroudis
Department of Computer Engineering and Informatics, University of Patras, 26504 Patras, Greece
e-mail: grivokwst@ceid.upatras.gr

I. Perikos
e-mail: perikos@ceid.upatras.gr

I. Hatzilygeroudis
e-mail: ihatz@ceid.upatras.gr

I. Hatzilygeroudis et al. (eds.), *Advances in Combining Intelligent Methods*, Intelligent Systems Reference Library 116, DOI 10.1007/978-3-319-46200-4_4

4.1 Introduction

The advent of the web has changed the way that educational tasks and learning processes are delivered to the students. It is a new platform that connects students with educational resources providing a number of advantages, like more efficient and personalized learning. Intelligent tutoring systems constitute a generation of computer-based educational systems that encompass intelligence to increase their effectiveness. Their main characteristic is that they adapt the educational task and learning activities to the individual student's needs, in order to maximize learning. This is mainly accomplished by utilizing Artificial Intelligence (AI) methods to represent the pedagogical decisions and the information regarding the domain to be taught, the learning activities and the student characteristics [1]. Educational systems, in order to better adapt to the learner, should provide proper educational tasks and leaning activities, tailored to the learner's knowledge level. The selection of the appropriate exercises to deliver to the student, according to the student's model and his/her knowledge level, is a central issue in intelligent e-learning [2, 3]. This selection usually takes into account two pieces of information, which are the student's knowledge level and also the difficulty level of the available exercises, and is mainly conducted by mimicking the corresponding human decision making.

The estimation of the difficulty level of exercises is of key importance in educational systems and assists in many vital educational processes. Good estimation of exercises difficulty level can result in better exercises sequencing, adapted to the student knowledge level. In this way, a tutoring system can offer more effective and personalized learning activities, maximizing student's comprehension [4]. Moreover, the assessment of the students can become more precise as well as the updates of the students' models and their knowledge levels, by taking difficulty level into account [5]. In most educational systems, the difficulty levels of the exercises are determined by the tutor when inserting them into the system. Such a determination of an exercise's difficulty level is a time-consuming task for the tutor and in some degree non-consistent. Several studies show that it is quite difficult for tutors to accurately estimate the difficulty level of questions or exercises in a consistent way [6]. Therefore, a system that can automatically estimate the difficulty level of an exercise in a consistent way could be a very useful support tool for a tutoring environment.

The determination of the difficulty level of an educational task is a very complex process. Tutors usually lack a formal reasoning mechanism to support their inference. In general, a tutor's exercise difficulty estimation is based on his/her experiences, sensitivities and standards. Of course, any tutor has intuitions about what aspects of an exercise and in what degree contribute in its complexity and difficulty. However, difficulty estimations by a tutor can be inconsistent and only approximations. Furthermore, every tutor has his/her personal subjective view of each exercise difficulty and if a number of tutors are asked to give precise definition of it, they will surely provide related but different statements [7].

In an artificial intelligence (AI) course, a fundamental topic is "search algorithms". It is considered necessary for students to get a strong understanding of the way those algorithms work and also of their implementation in solving various problems. However, many students have particular difficulties in understanding search algorithms. To assist students in learning search algorithms, we have developed and are using in our department the Artificial Intelligence Tutoring System (AITS). AITS [8, 9] is an intelligent tutoring system developed to help students in learning several topics of the artificial intelligence domain and also search algorithms. To this end, to assist students in learning search algorithms, a wide variety of interactive exercises and visualizations are offered. The system has been integrated into the artificial intelligence course curriculum in our department and offers students an interactive way to implement algorithms in several leaning scenarios and exercises. Students can study the theoretical aspects of the algorithms and learn to apply them through various interactive practice exercises. In order to tailor educational activities and exercises to the knowledge level and the performance of students, the difficulty level of the exercises needs to be accurately and consistently specified. However, AITS lacked a systematic approach and a mechanism to analyze the exercises and automatically determine their complexity and hence their difficulty level.

In this work, we present two approaches, a neuro-fuzzy [10] and a neurule-based one [11] for estimating the difficulty levels of exercises related to blind and heuristic search algorithms. The approaches aim to provide tutors with automatic ways to assess the difficulty levels of algorithmic exercises in a consistent and objective manner. To achieve that, initially, exercises on search algorithms are analyzed and characteristics or parameters that affect the complexity of the exercises are specified. Given that search algorithms mainly work on trees, such parameters are extracted from the topology representation of the tree and concerns: the number of the nodes, the average children that the nodes have, the depth of the tree and others. Then, two models, a neuro-fuzzy and a neurule-based one are trained based on exercise datasets created and annotated by expert tutors in agreement. The performances of the two models are quite satisfactory. The approaches adopted can help in representing tutor's experience in a way that can be interpreted and allows capturing tutor's subjectivity as well. The results collected from the experimental evaluation study are quite promising and show that both approaches are reliable, efficient and confirm the quality of their exercise difficulty identifications. This paper is an extension of [12].

The rest of the article is organized as follows: Sect. 4.2 presents the motivation of our work and describes the basic aspects of the exercise difficulty estimation area. Section 4.3 presents related work. Section 4.4 presents the neuro-fuzzy and the neurule approaches and illustrate their functionality. Section 4.5 presents the experimental study conducted and discusses the results collected. Finally, Sect. 4.6 concludes the article and provides directions for future work.

4.2 Motivation and Background

4.2.1 Motivation

Determination of the difficulty level of learning tasks or activities delivered to the students is very important and vital for maximizing learning efficiency and student understanding. In the literature, the methodologies used to determine the difficulty of educational tasks and exercises can be categorized in three main categories: statistical, heuristic and mathematical approaches. Statistical approaches utilize previous samples of students to estimate the difficulty level of an exercise. In heuristic approaches, the difficulty is determined directly by the human expert. Finally, mathematical approaches try to analyze the educational task and predict its difficulty based on its characteristics [7].

The estimation of the difficulty level of educational tasks and exercises is not an easy task and many studies question the ability of teachers to make consistent and accurate estimation of the difficulty level, since teachers usually fail to identify exercise's difficulty level in line with the exercises parameters and the students' ability [2, 13, 6]. Students could assist, in some degree, in determining exercises' difficulty. In some cases, it is reported that students estimate problem difficulty more accurate than tutors [14]. Additionally, tutors tend to misjudge students' performance. In general, they overestimate the performance of the students and underestimate the performance of borderline students [15]. In educational systems, it is considered necessary that the students should get exercises that are tailored to their learning needs and most of all are based on their knowledge level. The delivery of very easy educational activities to a student, lower than his/her level, can have no gain and cause learners lose any sense of learning and challenge. On the other hand, the delivery of very difficult and complex exercises, which are above his/her knowledge level and his/her previous performance, could frustrate and let down the student resulting in opposite learning effects [16].

So, for improving the educational process and the learning outcome, an intelligent education system should provide sequences of exercises and educational tasks to the students, tailored to their knowledge level. For the selection of the exercises that are proper to be delivered to the students, their difficulty levels should be estimated as accurately as possible.

4.2.2 Exercises on Search Algorithms

The Artificial Intelligence Tutoring System (AITS) offers theory about blind search (e.g. breadth-first, depth-first) and heuristic search (e.g. hill-climbing, branch-and-bound, A* etc.) algorithms, interactive examples and two types of exercises: practice exercises and assessment exercises. Practice exercises are interactive exercises that are equipped with hints and guidance during the learning sessions, aiming to help the student in learning AI search algorithms. On the other hand, assessment exercises are interactive exercises that are used to examine the student's progress and comprehension. An example of a simple interactive assessment exercise is illustrated in Fig. 4.1. In the context of an exercise, students are called to reach a specific goal node starting from the root (or an internal node of the tree) and following a requested AI search algorithm. Students can select a node to visit by clinking on it and the system provides guidance and feedback at student's actions during their interactions. Also, students can request for the system's assistance when stuck by clicking on the corresponding help button. AITS offer tests and exercises that are adapted to students needs in terms of knowledge level and performance. Adapted test can be generated and offered to students by the test generator unit of the system, which utilizes a rule-based expert system approach for making decisions on the difficulty level of the exercises to be included in the test [17], so that the test is adapted to the knowledge level and needs of the student. Created tests consist of a number of exercises that examine different aspects of search (blind and/or heuristic) algorithms. More specifically, a test on AI search algorithms consists of a number of interactive assessment exercises. For the adaptation of the exercises and the test to the students' needs, the accurate and consistent estimation of the exercises' difficulty level is needed. Due to the nature of the exercises on the domain of search algorithms, the difficulty estimation is based on the complexity of the graph or the tree representation of the exercise and also on characteristics of the exercise's solution.

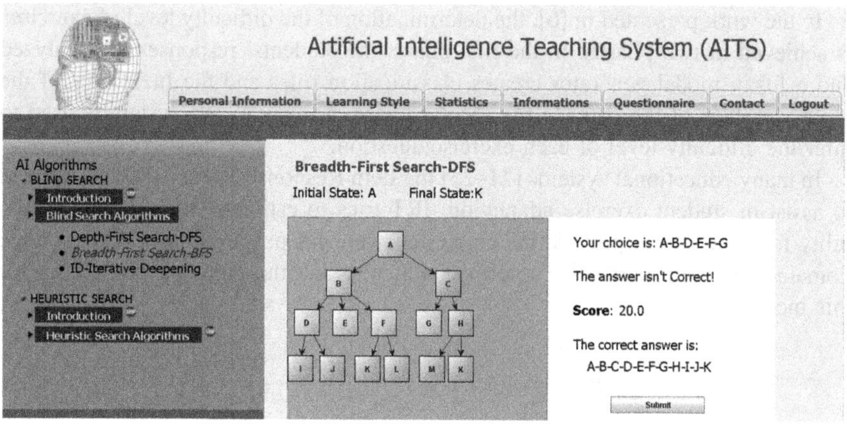

Fig. 4.1 An interactive assessment exercise on "breadth-first" search algorithm

4.3 Related Work

The accurate and consistent determination of the difficulty levels of exercises is of key importance in educational systems and can assist in many crucial educational processes. Several works study the determination of the difficulty of educational tasks in various educational domains. In the work presented in [18], authors present a hybrid AI approach which combines expert systems and genetic algorithm approaches for determining the difficulty level of exercises. A genetic algorithm is used to extract some kind of rules from the data acquired from the interactions of the students with the educational system, when solving exercises, and those rules are used to modify the expert rules provided by the tutor. In the work presented in [19], authors present a difficulty estimation approach for exercises related to logic, more specifically to formalizations of natural language sentences into first order logic formulas. An empirical approach is followed, where a difficulty estimating expert system is utilized to determine the difficulty level of exercises' conversion process based on a number of parameters related to the complexity of the process. Also, in the work presented in [20] authors present two different approaches for the determination of the difficulty of exercises on converting first order logic (FOL) formulas into clausal form (CF). The first approach is based on the complexity of the FOL formula, represented by a number of parameters, such as the number of atoms and connectives, while the second approach is based on the number and the complexity of the corresponding conversion steps needed for the exercise. The above approaches rely on the idea of analyzing both the exercise's remark and the exercise's answer in order to extract features and determine the difficulty level of the exercises.

In the work presented in [6], the determination of the difficulty level of exercises is achieved in two phases. In the first phase, the students' responses are analyzed and a fuzzy model generator creates classification rules and the fuzzy sets of the input variables of the data. In the second phase, a fuzzy expert system is used to infer the difficulty level of each exercise/question.

In many educational systems [21–23] the Item Response Theory (IRT) is utilized to assist in student exercise adaptation. IRT tries to estimate the guessing probability for a student to provide the correct answer, independently of the knowledge domain. This item parameter is termed difficulty and the guessing probability for this model is given by:

$$P_i(\theta) = c_i + \left(1 - c_i \frac{1}{1 + e^{-a_i(\theta - b_i)}}\right)$$

where, c_i is the guessing parameter for the probability that a student will answer correctly an exercise, e is the constant 2.718, b_i is the difficulty parameter, α_i represents the exercise discrimination parameter, stating how well the exercise can discriminate students of slightly different ability level, and θ is the learner's proficiency level. The variable b of an exercise difficulty level in most cases is denoted by the teacher. A system that implements IRT, in general, starts with exercises of medium difficulty. A correct student response will cause a more difficult to be offered next, while an incorrect response will cause a less difficult question to follow. So, the automatic estimation of the difficulty level of an exercise could be of great assistance in educational systems implementing IRT for the delivery of proper exercises to the student.

Hybrid methods that combine two or more intelligent methods to face complex problems have been applied to many aspects in the domain of intelligent educational systems. Neurules [24, 11] are a type of hybrid rules integrating symbolic rules with neurocomputing and can offer a number of benefits to ITSs, which are not offered by single ones. In the work presented in [17], authors present an ITS which employs neurules to make decisions during the teaching process. Expert knowledge is represented by neurules, which offer a number of benefits, such as time and space efficiency and reasoning robustness.

Another popular combination concerns the neuro-fuzzy approaches which have been used in various aspects of educational systems [25]. In general, neuro-fuzzy combinations can be implemented in two ways. A neural network can be equipped with the capability of handling fuzzy information or a fuzzy system can be augmented by neural networks to enhance some of its characteristics, like flexibility, speed, and adaptability. A representative of the latter combination is the well-known and widely used Adaptive Neuro-fuzzy Inference System (ANFIS) [10], which is utilized and examined in our work.

In the work presented in [26], authors present a neuro-fuzzy pedagogical recommender used to provide personalized learning experience by generating learning content recommendation based on students' knowledge and their learning style. In the work presented in [27], authors developed an adaptive neuro-fuzzy inference system to predict the student achievements in the engineering economy course. The aim of the system is to help students and evaluators to obtain more reliable and understandable results regarding students' performance. In the work presented in [28], authors present a neuro-fuzzy approach for student modeling. The authors monitor students' actions in an exploratory learning environment and collect best available information for students' diagnosis and assess their knowledge level. Then utilize a neuro-fuzzy approach to classify students into different knowledge level categories and report better performance than other methods, such as pure neural-networks. In the work presented in [29], authors developed a fuzzy rule system that consists of a set of 18 fuzzy rules, used in assessing student performance and learning efficiency, obtained from experts. For the assessment, four input parameters are taken into consideration, which are the average marks, the time spent, the number of student's attempts and the number of help requests. The combination with the neural network leads to the application of an ANFIS implementation, which reports quite promising performance. In the work presented in [30], authors propose a student classification technique based on an adaptive neuro-fuzzy system. They report that the adaptive neuro-fuzzy system fits in modeling the field of control systems, expert systems and complex systems and in the area of educational systems. The authors' motivation behind their research work is to accurately identify exceptional and weak students for proper attention. In the work presented in [31], authors present a neuro-fuzzy approach for classifying students into different groups. The neuro-fuzzy classifier is based on previous exam results and student related factors, such as gender, as input variables and labels students based on their expected overall academic performance. The results showed that the proposed approach achieved a high accuracy and outperforms other widely used methods, such as neural networks, support vector machines, Naïve Bayes and decision trees.

In this work, we study the performance of the neurule-based and neuro-fuzzy approaches in determining the difficulty level of exercises on the domain of search algorithms. Due to the high complexity and dimensionality of the domain's concepts and the tree representation of the exercises, hybrid and neural approaches seem to be quite suitable in capturing the intuition and subjectivity of expert's and the work presented in this article is a contribution towards examining this direction.

4.4 Neuro-Fuzzy and Neurule-Based Approaches for Exercise Difficulty Estimation

In this section, we present the neuro-fuzzy and the neurule approaches, describe their principles and analyze their functionality. Initially, we illustrate the way that exercises are analyzed and proper features are extracted. For the estimation of the difficulty of exercises, the tree topology representation of the exercises is analyzed and a set of features are specified from there.

4.4.1 Exercise Analysis and Feature Extraction

Initially, the exercises are analyzed, proper features are extracted and their corresponding numerical parameters are calculated. Based on our cooperation with experts, we specified the set of parameters that affect and contribute in the difficulty level of the exercises. So, we resulted in a number of parameters, which can be distinguished in:

- Input parameters: The number of the nodes, the average number of the children of each node, the maximum depth of the tree, the length of the solution concerning the nodes needed to visit in order to reach the goal node.
- Output parameters: The difficulty level of tree exercise which can take the following values: *very easy*, *easy*, *medium*, *difficult* and *advanced.*

The input parameters and their value range are depicted in the following (Table 4.1).

Table 4.1 Input variables

Input variable	Range
Number of nodes	3–40
Average number of the children	2–8
Depth of the tree	3–12
Solution's length	1–40

Each one of the parameters that are presented above captures and conveys different aspects of the complexity of an exercise. The number of nodes represents the total nodes that the tree graph representation of an exercise consists of. In the context of our experimental study, the test exercises that were used, consisted of up to 40 nodes. The average number of the children, which the nodes of the tree structure have, represents the possible expansion choices that can be made by the students at each node of the exercise. This variable can take continuous values ranging from 2 to 8. The depth of the tree represents the number of the levels of the graph tree exercises and can take integer values ranging from 3 to 12. This parameter is again related to the complexity of the tree graph; the larger the number the more complex the tree hence the more difficult the exercise.

Finally, the solution length parameter represents the number of nodes that must be visited in order to reach the goal node from the start node. This parameter takes integer values and the highest value it can take is the maximum number of the graph's nodes which is set to 40. The parameter represents the steps that the student has to make in order to complete the exercise, and in line with the other parameters, can provide meaningful indicators of the complexity level of the exercises.

4.4.2 Neuro Fuzzy Approach

The neuro-fuzzy difficulty estimation mechanism takes as input specific characteristics of the exercises and provides as output the difficulty level of the exercises. In the ANFIS approach, fuzzy logic is used to capture the subjectivity of the teacher and the linguistic description of the exercises parameters. In general, the neuro-fuzzy model consists of 5 main layers as presented in Fig. 4.2.

In our case, Layer 1, which is the input layer of the neuro-fuzzy model, has 4 neurons, which represent the four input variables. The input layer sends out external crisp values directly to the next layer (Layer 2), which is the fuzzification layer, and the crisp values from the first layer are transformed to the appropriate linguistic fuzzy values (low, average or high). The neurons of Layer 2 represent a fuzzy set for each one of the input linguistic variables. The output of the layer is the degree of membership of each input value for each fuzzy set. Each node output is considered to represent the firing strength of a rule. The neurons of Layer 3 (fuzzy rule layer) represent fuzzy rules (R1, R2, …, Rk) and each node calculates the ratio of the associated rules' firing strength. The outputs of this layer are called normalized firing strengths. Layer 4 (output membership layer) consists of adaptive nodes,

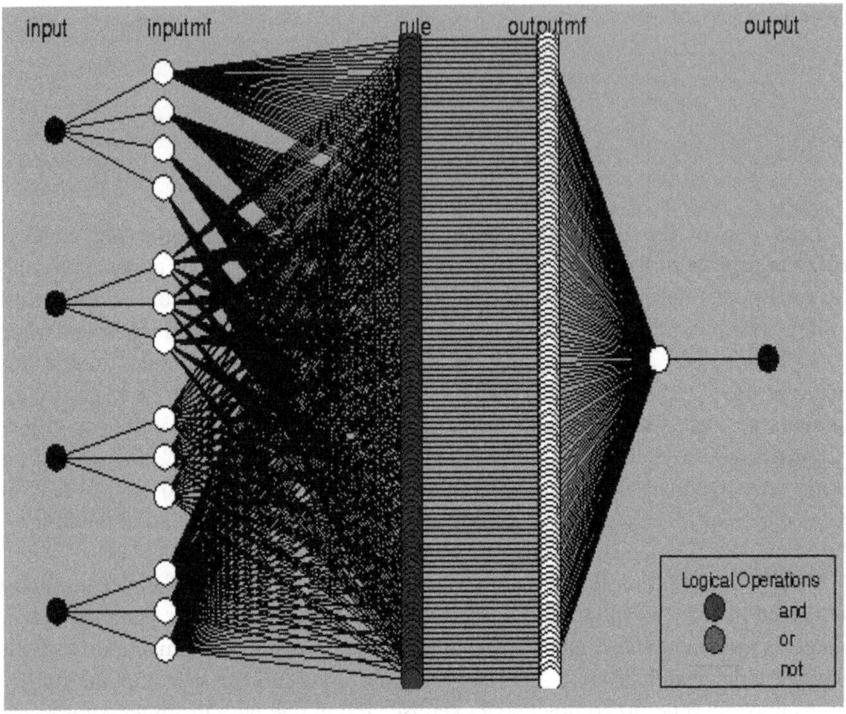

Fig. 4.2 The basic architecture of the neuro-fuzzy model

which take inputs from the fuzzy rule neurons and merge them by using fuzzy union operations. Parameters in this layer are referred to as consequent parameters. Layer 5 (defuzzification layer) consists of a single fixed node labeled Σ, which computes the output of our system as the sum-product composition of all incoming signals.

In Fig. 4.3, the membership functions for the input parameter representing the number of tree graph's nodes are presented. The input parameter can take four fuzzy sets (linguistic values) which are: "very-few", "few", "medium" and "many" are presented.

Training ANFIS is based on a hybrid algorithm combining the widely used least-squares and the gradient descent techniques. During the forward pass of the hybrid learning algorithm, the outputs of the nodes go forward until layer 4 and the consequent parameters are identified by the least-squares method. During the backward pass, the error signals propagate backwards and the premise parameters are updated by gradient descent. The error measure to train the above-mentioned ANFIS is the following:

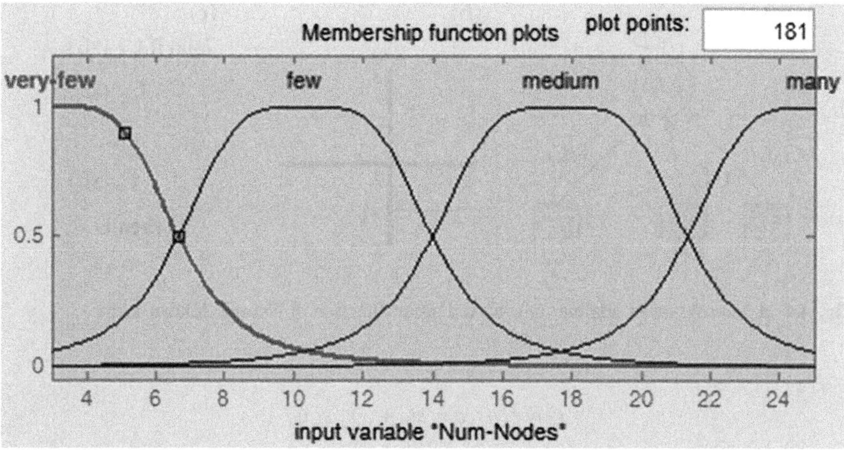

Fig. 4.3 Membership functions for the number of nodes parameter

$$E = \sum_{k=1}^{n} \left(f_k - \widehat{f_k}\right)^2$$

where f_k and $\widehat{f_k}$ are the *k*th desired and estimated output respectively, and *n* is the total number of pairs (inputs–outputs) of data in the training set. The output error is used to adapt the premise parameters used, utilizing the back propagation algorithm. This hybrid algorithm is proven to be very efficient in training ANFIS models [10, 32]. In the context of our study, the ANFIS model was trained based on the dataset of exercises that were annotated by the experts in agreement. After the training of neuro-fuzzy model is completed, it can be used to classify new exercises into the appropriate difficulty category.

4.4.3 Neurule-Based Approach

Neurules (**Neu**ral **rules**) are a kind of hybrid rules. Each neurule (Fig. 4.4a) is considered as an adaline unit (Fig. 4.4b). The inputs C_i (i = 1,…, n) of the unit are the conditions of the rule. Each condition C_i is assigned a number sf_i, called a significance factor, corresponding to the weight of the corresponding input of the adaline unit. Moreover, each rule itself is assigned a number sf_0, called the bias factor, corresponding to the bias of the unit. Each input takes a value from the following set of discrete values: [1 (true), −1 (false), 0 (unknown)]. The output *D*, which represents the conclusion (decision) of the rule, is calculated via the formulas:

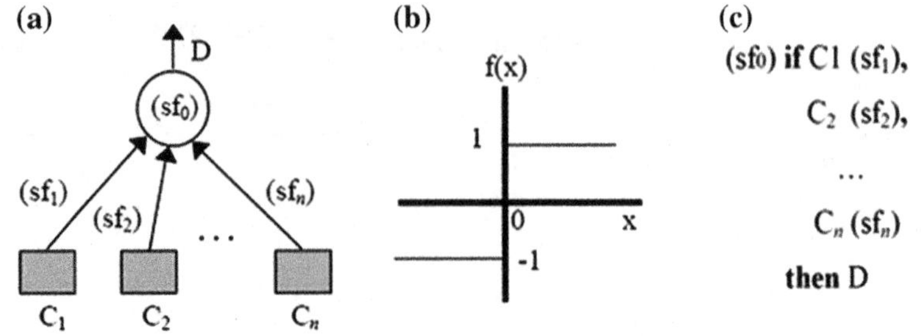

Fig. 4.4 **a** Neurule as an adaline unit. **b** Activation function. **c** Neurule textual form

$$D = f(a), a = sf_0 + \sum_{i=1}^{n} sf_i C_i$$

where **a** is the *activation value* and *f(x)* the *activation function*, a threshold function:

$$f(a) = \begin{cases} 1 & if\ a \geq 0 \\ -1 & otherwise \end{cases}$$

Hence, the output can take one of two values, '−1' and '1', representing failure and success of the rule respectively.

The textual form of a neurule is presented in Fig. 4.4c. C_is are the *conditions* and *D* represents the *conclusion* (decision) of the rule. The significance factor sf_i of a condition represents the significance (weight) of the condition in drawing the conclusion.

Neurules can be produced from empirical data through a well-defined process [33]. The produced neurules constitute an integrated rule base, which can be used for making inferences [33] and even produce explanations [34], through a well-defined inference process [33].

4.5 Experimental Evaluation

We conducted an experimental study of the two approaches with the aim to evaluate their performance. For the evaluation study, we used a dataset of 240 different exercises on search algorithms. For the needs of the study, two expert tutors teaching the course of artificial intelligence and having many years of experience, were asked to assist in the difficulty annotation of the corpus of the exercises. For each one, the tutors specified in agreement the proper difficulty level of the exercise and after that, the dataset was created. The exercises of the dataset were collected from textbooks and web resources and also created for the needs of this study. In the dataset, each row includes values of the difficulty parameters of the exercise and the corresponding difficulty level, annotated by the experts. For the needs of this study, the two thirds of the dataset were used for training, i.e. for producing the models of the two approaches, and the remaining one third was used for the evaluation of the models.

Given that we have a multiple class output, we use the following four metrics: average accuracy, precision, recall and F-measure. Average accuracy is the mean value of the accuracies of the output classes and is calculated as follows:

$$Average\,Accuracy = \frac{\sum_{i=1}^{l} \frac{tp_i + tn_i}{tp_i + fn_i + fp_i + tn_i}}{l}$$

where *tp*, *fn*, *fp* and *tn* correspond to true positives, false negatives, false positives and true negatives respectively. F-measure is the harmonic mean of precision and recall and is defined as:

$$F - measure = 2 \times \frac{prec \times rec}{prec + rec}$$

Initially, we present the performance results of the neuro-fuzzy approach. The evaluation results of the performance of the ANFIS model classifier are presented in Table 4.2. Evaluation refers to comparing the outputs of the neuro-fuzzy system with the difficulty levels specified by the expert-tutors.

The results indicate that the mechanism has a very encouraging performance. Approximately in 85 % of the cases the neuro-fuzzy model was able to specify the correct level of difficulty of the exercises. Also, the performance results of the neurules model are presented in Table 4.3.

Both models showed a very good performance. The best performance is achieved by the neuro-fuzzy model, which accuracy and precision are better than those of the neurule approach in the experimental study. Also, a noticeable point of both models is that they have a very good precision which is 0.84 for both of them. The fact that the models do not report 100 % accuracy, compared to the experts, it is not actually a failure, given that experts have problems in accurately estimating

Table 4.2 Evaluation results of the neuro-fuzzy model

Average accuracy	0.85
Precision	0.84
Recall	0.85
F-measure	0.84

Table 4.3 Evaluation results of the neurules model

Average accuracy	0.78
Precision	0.84
Recall	0.78
F-measure	0.81

Table 4.4 Cohen's kappa metric for neuro-fuzzy and neurule approaches

	Neuro-fuzzy	Neurules
Expert human rater	0.81	0.72

difficulty levels of exercises in a consistent way. Also, they couldn't consistently explain the found differences.

In order to get a deeper insight of the performance of the approaches and their agreement with the expert tutors, we calculate the Cohen's Kappa statistic [35], which is defined as follows:

$$k = \frac{p_0 - p_e}{1.0 - p_e}$$

where p_o represents the proportion of rater exhibiting agreement and p_e represents the proportion expected to exhibit agreement by chance alone. Thus, "perfect agreement" would be indicated by "$k = 1$" and no agreement means that "$k = 0$". In general, a value of kappa higher than 0.8 can indicate an almost perfect agreement, whereas a value between 0.61 to 0.8 can indicate a significant agreement [36]. In the context of this study, the Cohen's Kappa metric was adopted and was calculated into order to specify if there was any agreement between the difficulty estimations of the tutors and the difficulty estimations of the two approaches. The results collected are presented in Table 4.4.

The results indicate that there was an almost perfect agreement between the expert tutors and neuro-fuzzy approach, given that the metric was calculated to be $k = 0.81$ (95 % confidence interval, $p < 0.0005$). In addition, the agreement between the experts and the neurule approach was slightly lower and the Kappa metric was calculated to be $k = 0.72$ (95 % confidence interval, $p < 0.0005$), which indicates a very good agreement.

4.6 Conclusions

In this article we address the issue of automatically determining the difficulty level of search algorithm exercises. The aim is to provide the tutors with an automatic way to assess the difficulty level of algorithmic exercises in a consisted and objective manner as well as to reduce their workload. The estimation of the exercises difficulty, when made in a systematic way, can save considerable time for tutors, especially when a large corpus of exercises is required. Moreover, accuracy and consistency are usually difficult to be achieved by tutors, due to subjective reasons and therefore a system that can automatically, consistently and objectively determine the difficulty level of exercises is of great help.

In this work, we utilized a neuro-fuzzy and a neurule based approaches to learn to classify exercises into the proper difficulty categories. Initially, exercises on search algorithms were analyzed and proper features were extracted from their tree representation. The neuro-fuzzy approach helps to represent tutor's experience in a way that can be interpreted and allows capturing tutor's subjectivity. At each part/parameter of an exercise a linguistic term is annotated, combining fuzzy evidences, which contributes in some degree to the overall difficulty/complexity level of the exercise. Fuzzy logic and linguistic terms can be awarded to a single exercise's characteristic as well as to a group of exercises which have similar characteristics and parameters. The neurule-based approach is also quite suitable. It integrates symbolic rules with neurocomputing and can offer a number of benefits such as time and space efficiency and reasoning robustness.

The evaluation experiments conducted and the results gathered indicate quite satisfactory performance. Both approaches achieved quite satisfactory performance where the neuro-fuzzy approach reported a better performance compared to the neurules. In addition, the experts involved in the evaluation study stated that they do not usually follow such a systematic and analytical way of estimating exercises' difficulty and that they saw the creation of such a system as an opportunity to express in a systematic way their views and experiences. On the other hand, they admitted that they are not always consistent, mainly due to the fact that they usually act intuitively. Given the above, they also admitted that the automatic approaches examined and formulated in this work, can do better than them in some cases due to their consistency.

However, there are directions that future work could focus on. Initially, conducting a larger scale experimental evaluation and evaluating the system on a larger number of interactive exercises on algorithms will give us a deeper insight of the performances of the two approaches. Also, neurules can improve their performance by incorporating case-based reasoning [24]. Applying this capability is another direction for further research. In addition, currently the hybrid approaches presented and examined in this work can estimate the difficulty of tree-based algorithm exercises based on the analysis and the features of the tree representation of the exercises. An extension will be the analysis and the difficulty estimation of graph-based exercises, where search algorithms are to be applied on complex graph networks. Finally, another direction for future work concerns the examination of ensemble classifier schemas that combine base classifiers under different ensemble approaches. Exploring this direction is a key aspect of our future work.

Acknowledgment This work was partially supported by the Research Committee of the University of Patras, Greece, Program "Karatheodoris", project No C901.

References

1. Hatzilygeroudis, I., Giannoulis, C., Koutsojannis, C.: Combining expert systems and adaptive hypermedia technologies in a web based educational system. In: Fifth IEEE International Conference on Advanced Learning Technologies, ICALT 2005, pp. 249–253. IEEE (2005)
2. Alexandrou-Leonidou, V., Philippou, G.N.: Teachers' beliefs About Students Development of the Pre-algebraic Concept of Equation. International Group for the Psychology of Mathematics Education, p. 41 (2005)

3. Barla, M., Bieliková, M., Ezzeddinne, A.B., Kramár, T., Šimko, M., Vozár, O.: On the impact of adaptive test question selection for learning efficiency. Comput. Educ. **55**(2), 846–857 (2010)
4. Chen, L.H.: Enhancement of student learning performance using personalized diagnosis and remedial learning system. Comput. Educ. **56**(1), 289–299 (2011)
5. Nguyen, M.L., Hui, S.C., Fong, A.C.M.: Large-scale multiobjective static test generation for web-based testing with integer programming. IEEE Trans. Learn. Technol. **6**(1), 46–59 (2013)
6. Verdú, E., Verdú, M.J., Regueras, L.M., de Castro, J.P., García, R.: A genetic fuzzy expert system for automatic question classification in a competitive learning environment. Expert Syst. Appl. **39**(8), 7471–7478 (2012)
7. Conejo, R., Guzmán, E., Perez-De-La-Cruz, J.L., Barros, B.: An empirical study on the quantitative notion of task difficulty. Expert Syst. Appl. **41**(2), 594–606 (2014)
8. Grivokostopoulou, F., Hatzilygeroudis, I.: Teaching AI search algorithms. In: A Web-Based Educational System, Proceedings of the IADIS International Conference e-Learning 2013, 23–26 July, pp. 83–90. Prague, Czech Republic (2013)
9. Grivokostopoulou, F., Perikos, I., Hatzilygeroudis, I.: An educational system for learning search algorithms and automatically assessing student performance. Int. J. Artif. Intell. Educ. (2016). doi:10.1007/s40593-016-0116-x
10. Jang, J.S.: ANFIS: adaptive-network-based fuzzy inference system. Syst. Man Cybern. IEEE Trans. **23**(3), 665–685 (1993)
11. Hatzilygeroudis, I., Prentzas, J.: Constructing modular hybrid knowledge bases for expert systems. Int. J. Artif. Intell. Tools **10**, 87–105 (2001)
12. Grivokostopoulou, F., Perikos, I., & Hatzilygeroudis, I. (2015). Estimating the Difficulty of Exercises on Search Algorithms Using a Neuro-fuzzy Approach. In *Tools with Artificial Intelligence (ICTAI), 2015 IEEE 27th International Conference on* (pp. 866–872). IEEE
13. Van de Watering, G., Van der Rijt, J.: Teachers' and students' perceptions of assessments: a review and a study into the ability and accuracy of estimating the difficulty levels of assessment items. Educ. Res. Rev. **1**(2), 133–147 (2006)
14. Lee, F.L.: Electronic homework: an intelligent tutoring system in mathematics. Doctoral Dissertation, The Chinese University of Hong Kong Graduate School—Division of Education, November (1996). http://www.fed.cuhk.edu.hk/en/cuphd/96fllee/content.htm. Assessed June 2014
15. Impara, J.C., Plake, B.S.: Teachers' ability to estimate item difficulty: a test of the assumptions in the Angoff standard setting method. J. Educ. Meas. **35**(1), 69–81 (1998)
16. Leung, E.W.C., Li, Q.: An experimental study of a personalized learning environment through open-source software tools. IEEE Trans. Educ. **50**(4), 331–337 (2007)
17. Hatzilygeroudis, I., Prentzas, J.: Using a hybrid rule-based approach in developing an intelligent tutoring system with knowledge acquisition and update capabilities. Expert Syst. Appl. **26**(4), 477–492 (2004)
18. Koutsojannis, C., Beligiannis, G., Hatzilygeroudis, I., Papavlasopoulos, C.: Using a hybrid AI approach for exercise difficulty level adaptation. Int. J. Continuing Eng. Educ. Life Long Learn. **17**(4), 256–272 (2007)
19. Perikos, I., Grivokostopoulou, F., Hatzilygeroudis, I., Kovas, K.: Difficulty estimator for converting natural language into first order logic. In: Intelligent Decision Technologies, pp. 135–144. Springer, Berlin, Heidelberg (2011)
20. Grivokostopoulou, F., Hatzilygeroudis, I., Perikos, I.: Teaching assistance and automatic difficulty estimation in converting first order logic to clause form. Artif. Intell. Rev. 1–21 (2013)
21. Chen, C.M., Duh, L.J.: Personalized web-based tutoring system based on fuzzy item response theory. Expert Syst. Appl. **34**(4), 2298–2315 (2008)
22. Cheng, I., Shen, R., Basu, A.: An algorithm for automatic difficulty level estimation ofmultimedia mathematical test items. Proceedings of the 8th IEEE International Conference

on Advanced Learning Technologies, pp. 175–179. IEEE Computer Society, Los Alamitos, CA (2008)
23. Wauters, K., Desmet, P., Van Den Noortgate, W.: Acquiring item difficulty estimates: a collaborative effort of data and judgment. In: Education Data Mining (2011)
24. Hatzilygeroudis, I., Prentzas, J.: Integrating (rules, neural networks) and cases for knowledge representation and reasoning in expert systems. Expert Syst. Appl. **27**(1), 63–75 (2004)
25. Kar, S., Das, S., Ghosh, P.K.: Applications of neuro fuzzy systems: a brief review and future outline. Appl. Soft Comput. **15**, 243–259 (2014)
26. Sevarac, Z., Devedzic, V., Jovanovic, J.: Adaptive neuro-fuzzy pedagogical recommender. Expert Syst. Appl. **39**(10), 9797–9806 (2012)
27. Taylan, O., Karagözoğlu, B.: An adaptive neuro-fuzzy model for prediction of student's academic performance. Comput. Ind. Eng. **57**(3), 732–741 (2009)
28. Stathacopoulou, R., Grigoriadou, M., Magoulas, G.D., Mitropoulos, D.: A neuro-fuzzy approach in student modeling. In: User Modeling 2003, pp. 337–341. Springer, Berlin, Heidelberg (2003)
29. Norazah, Y.: Student learning assessment model using hybrid method. Universiti Keba, PhD Thesis (2005)
30. Iraji, M.S., Aboutalebi, M., Seyedaghaee, N.R., Tosinia, A.: Students classification with adaptive neuro fuzzy. Int. J. Mod. Educ. Comput. Sci. (IJMECS) **4**(7), 42 (2012)
31. Do, Q.H., Chen, J.F.: A neuro-fuzzy approach in the classification of students' academic performance. Comput. Intell. Neurosci. **2013**, 6 (2013)
32. Kakar, M., Nyström, H., Aarup, L.R., Nøttrup, T.J., Olsen, D.R.: Respiratory motion prediction by using the adaptive neuro fuzzy inference system (ANFIS). Phys. Med. Biol. **50** (19), 4721 (2005)
33. Hatzilygeroudis, I., Prentzas, J.: Integrated rule-based learning and inference. IEEE Trans. Knowl. Data Eng. **22**, 1549–1562 (2010)
34. Hatzilygeroudis, I., Prentzas, J.: Symbolic-neural rule based reasoning and explanation. Expert Syst. Appl. **42**, 4595–4609 (2015)
35. Cohen, J.: A coefficient of agreement for nominal scales. Educ. Psychol. Measur. **20**(1), 37–46 (1960)
36. Viera, A.J., Garrett, J.M.: Understanding interobserver agreement: the kappa statistic. Fam. Med. **37**(5), 360–363 (2005)

Chapter 5
Generation and Nonlinear Mapping of Reducts—Nearest Neighbor Classification

Naohiro Ishii, Ippei Torii, Kazunori Iwata and Toyoshiro Nakashima

Abstract Dimension reduction of data is an important theme in data processing. Reduct in the rough set is useful since it has the same discernible power as the entire features in the higher dimensional scheme. But, classification with higher accuracy is not obtained in the reduct followed by nearest neighbor processing. To deal with the problem, it is shown that nearest neighbor relation with minimal distance introduced here has a basic piece of information for classification. In this paper, a new reduct generation method based on the nearest neighbor relation with minimal distance is proposed. To improve the classification accuracy of reducts, we develop a nonlinear mapping and embedding methods on the nearest neighbor relation, which also adjust vector data relation and preserve data ordering to cope with noise in classification.

Keywords Classification · Reduct generation · Nearest neighbor relation with minimal distance · Nonlinear mapping and embedding

---

N. Ishii (✉) · I. Torii
Aichi Institute of Technology, Toyota, Japan
e-mail: ishii@aitech.ac.jp; nishii@acm.org

I. Torii
e-mail: mac@aitech.ac.jp

K. Iwata
Aichi University, Nagoya, Japan
e-mail: kazunori@vega.aichi-u.ac.jp

T. Nakashima
Sugiyama Jyogakuen University, Nagoya, Japan
e-mail: nakasima@sugiyama-u.ac.jp

I. Hatzilygeroudis et al. (eds.), *Advances in Combining Intelligent Methods*,
Intelligent Systems Reference Library 116, DOI 10.1007/978-3-319-46200-4_5

5.1 Introduction

Rough sets theory firstly introduced by Pawlak [1–4] provides us a new approach to perform data analysis, practically. Up to now, rough set has been applied successfully and widely in machine learning and data mining. The need to manipulate higher dimensional data in the Web and to support or process them gives rise to the question of how to represent the data in a lower-dimensional space to save space and computation time. Thus, dimension reduction of data is an important problem [5–7]. An important task in rough set based data analysis is computation of the attributes or feature reducts for the classification. By Pawlak's [1–4, 8, 9] rough set theory, a reduct is a minimal subset of features, which has the discernibility power as using the entire features. For the classification, nearest neighbor method [10] is simple and effective one. We have problems for the application of the nearest neighbor method to reducts classification [11–14]. Nearest neighbor relation with minimal distance between different classes is proposed here as a basic information for classification. We propose here a new reduct generation based on the nearest neighbor relation with minimal distance. To improve the classification accuracy of the reduct, firstly a classification based on the linearly separable condition is developed from convex hull. Then, the condition developed here is reduced to the construction of independent data vectors. For the independence of data vectors, the reduct is mapped to higher dimensional space, which is realized by the product of variables in a nonlinear mapping. The nonlinear mapping is based on the nearest neighbor relation. It is shown that nonlinear mapping and embedding of the reduct shows the classification with higher accuracy. Further, data classification is realized in the vector operations based on the nearest neighbor relation.

5.2 Generation of Reducts Based on Nearest Neighbor Relation

In [8, 9] decision table is represented as a discernibility matrix. This representation has some advantages, since data dimensional reduction can be realized on the discernibility matrix [1–4, 8]. The discernibility matrix is defined as follows.

Definition 5.1 An information system is composed of a 4-tuple as $T = \{U, A, C, D\}$ where U is a finite nonempty set of n instances, $\{X_1, X_2, \ldots, X_n\}$. A is a finite set of attributes. C is a set of attribute value and D is a set of decision function. By a discernibility matrix of T, denoted by $M(T)$, which is $n \times n$ matrix as

$$\begin{aligned} m_{ij} &= \{(\alpha \in C{:}\, \alpha(X_i) \neq \alpha(X_j)) \wedge (\beta \in D, \beta(X_i) \neq \beta(X_i))\} \\ i,j &= 1, 2, \ldots n \end{aligned} \tag{5.1}$$

where α denotes a decision function of instance and β denotes decision function of the class. This paper aims to newly develop a generation method of reducts based on the nearest neighbor relation proposed here. A simple example of a decision table is shown in Table 5.1. In Table 5.1, a, b, c and d are attributes for instances $\{X_1, X_2,\ldots, X_7\}$ with two classes +1 and −1. From Table 5.1, the discernibility matrix is generated as shown in Table 5.2, which is made for the differences among instances [1–4, 8]. We can define a new concept, a nearest neighbor relation with minimal distance, δ. Instances with different classes are assumed to be measured in the metric distances for the nearest neighbor classification.

Table 5.1 Data example of decision table

Attrib. / Inst.	a	b	c	d	class
X_1	1	0	2	1	+1
X_2	1	0	2	0	+1
X_3	2	2	0	0	−1
X_4	1	2	2	1	−1
X_5	2	1	0	1	−1
X_6	2	1	1	0	+1
X_7	2	1	2	1	−1

Table 5.2 Discernibility matrix of the decision table

Inst. / Inst.	X_1	X_2	X_3	X_4	X_5	X_6
X_1	—					
X_3	a,b,c,d	a,b,c				
X_4	b	b,d	—			
X_5	a,b,c	a,b,c,d	—	—		
X_6	—	—	b,c	a,b,c,d	c,d	
X_7	a,b	a,b,d	—	—	—	c,d

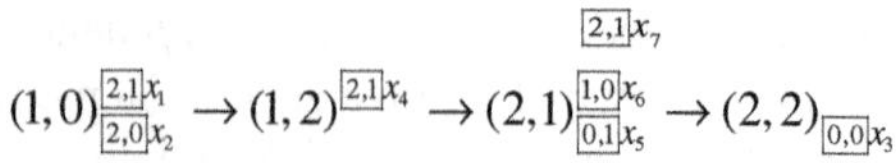

Fig. 5.1 Lexicographical ordering for nearest neighbor relation

Definition 5.2 A nearest neighbor relation with minimal distance is a set of pair of instances, which are described in,

$$\{(X_i, X_j): \beta(X_i) \neq \beta(X_j) \wedge |X_i - X_j| \leq \delta\} \tag{5.2}$$

where $|X_i - X_j|$ in Eq. (5.2) denotes the distance between X_i and X_j, and δ is the minimal distance.

When Eq. (5.2) holds for the given $\delta > 0$, X_i and X_j are called to be in the nearest neighbor relation with minimal distance δ. This nearest neighbor relation plays a fundamental role for the generation of reducts from the discernibility matrix. In the Table 5.1, to find the nearest neighbor relation with minimal distance δ, a lexicographical ordering for the instances is developed as shown in Fig. 5.1, where the values of the first two attributes are shown within the parentheses and the values of the last two attributes are shown within a box. In Fig. 5.1, the Euclidean distance between X_6 (2110) in class +1 and X_7 (2121) in class −1, becomes $\sqrt{2}$. Similarly, the distances between X_5 (2101) and X_6 (2110), between X_1 (0021) and X_7 (2121), and between X_3 (2200) and X_6 (2110) are $\sqrt{2}$.

In Fig. 5.1, (X_6, X_7), (X_5, X_6), (X_1, X_7) and (X_3, X_6) are elements of the relation with a distance $\sqrt{2}$. Thus, a nearest neighbor relation with minimal distance $\sqrt{2}$ becomes

$$\{(X_1, X_7),\ (X_5, X_6),\ (X_6, X_7),\ (X_3, X_6)\} \tag{5.3}$$

5.2.1 *Generation of Reducts Based on Nearest Neighbor Relation with Minimal Distance*

Here, we want to introduce the nearest neighbor relation on the discernibility matrix. Assume that the set of elements of the nearest neighbor relation are $\{nn_{ij}\}$.

Then, the following characteristics are shown. Respective element of the set $\{nn_{ij}\}$ corresponds to the term of Boolean sum. As an example, the element {a, b, c} of discernibility matrix in the set $\{nn_{ij}\}$ corresponds to a Boolean sum (a + b + c). The following lemmas are derived easily.

Lemma 5.1 *Respective Boolean term consisting of the set* $\{nn_{ij}\}$ *becomes a necessary condition to be reducts in the Boolean expression.*

This is trivial, since the product of respective Boolean term becomes reducts in the Boolean expression.

Lemma 5.2 *Boolean product of respective terms corresponding to the set* $\{nn_{ij}\}$ *becomes a necessary condition to be reducts in the Boolean expression.*

This is also trivial by the reason of Lemma 5.1. Thus, the relation between Lemmas 5.1 and 5.2 is described as follows.

Lemma 5.3 *Reducts in the Boolean expression are included in the Boolean term of Lemma 5.1 and the Boolean product in Lemma 5.2.*

Figure 5.2 shows that nearest neighbor relation with classification is a necessary condition in the Boolean expression for reducts, but not sufficient condition. The distance δ of the nearest neighbor relation in the Definition 5.2 is compared with the distance δ' of the relation in the following theorem.

Theorem 5.1 *If the distance* δ *is greater than the* δ', *i.e.,* $\delta > \delta'$ *in the Definition 5.2, the Boolean expression of the case of* δ' *includes that of* δ.

This is by the reason that the Boolean expression of the nearest neighbor relation consists of the Boolean product of variables of the relation. The number of variables in the distance δ' are less than that of δ. Thus, the nearest neighbor relation with distance δ' includes the ellipse of δ in Fig. 5.2.

Two classes of attributes (variables), (A) and (B) are defined to extract reducts in the nearest neighbor relation $\{nn_{ij}\}$.

(A): Attributes(variables) in the discernibility matrix includes those of any respective element in $\{nn_{ij}\}$.

(B): Attributes(variables) in the discernibility matrix are different from those of any respective element in $\{nn_{ij}\}$. The elements in class (A) are absorbed in the corresponding elements of the set $\{nn_{ij}\}$ in the Boolean expression.

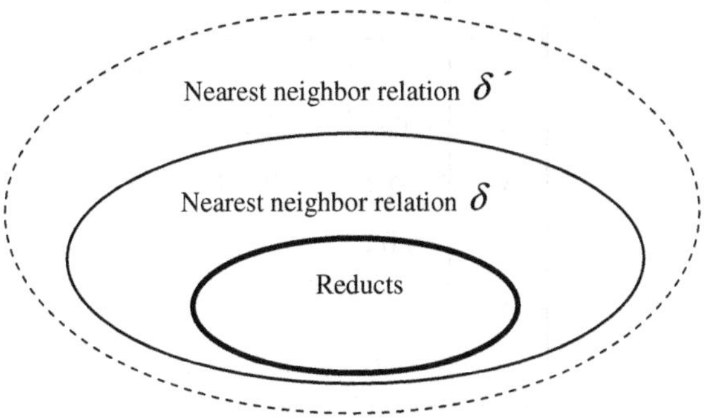

Fig. 5.2 Boolean condition of nearest neighbor relations and reducts

The Boolean sum of attributes is absorbed in the Boolean sum element with the same fewer attributes in the set $\{nn_{ij}\}$. As an example, the Boolean sum of (a + b + c) in the class (A) is absorbed in the Boolean sum of (a + b) in the set $\{nn_{ij}\}$.

Lemma 5.4 *Within class (B), the element with fewer attributes (variables) plays a role of the absorption for the element with more attributes (variables). Then, the absorption operation is carried out within class (B).*

Theorem 5.2 *Reducts are derived from the nearest neighbor relation with the absorption of class (B) terms in the Boolean expression.*

In the relation (X_1, X_7) in Table 5.1, the X_1 in the class +1 is nearest to the is X_7 in the class −1. Similarly, (X_5, X_6), (X_6, X_7) and (X_3, X_6) are nearest relations. Then, nearest neighbor relations in class (A) are shown in shading cell in Table 5.2 of the discernible matrix. Similarly, instances of these relations are shown in shading cells in Table 5.1. The Boolean product of these four terms becomes,

$$(a+b)\cdot(b+c)\cdot(c+d)=b\cdot c+b\cdot d+a\cdot c \quad (5.4)$$

which becomes a candidate of reducts. The third term in Eq. (5.4) is absorbed by the product of variable {b} of the class (B) and Eq. (5.4). The final reducts equation becomes

$$b\cdot c+b\cdot d \quad (5.5)$$

Thus, reducts $b\cdot c$ and $b\cdot d$ are obtained.

As another example, let the attribute b of the instance x_3 be changed to 1 and the attribute c of the instance x_4 be changed to 1 in the values in Table 5.1. Then, the discernibility matrix in Table 5.2 is changed to that in Table 5.3.

In Table 5.3, the shadowed elements show the nearest neighbor with minimal distance, $\sqrt{2}$. From these elements, the nearest neighbor relation is represented as

Table 5.3 Dicernibility matrix in modified decision table

Inst. \ Inst.	X_1	X_2	X_3	X_4	X_5	X_6
X_2	—					
X_3	a,b,c,d	a,b,c				
X_4	b,c	a,b,c,d	—			
X_5	a,b,c	a,c,d	—	—		
X_6	—	—	a,c	b,d	c,d	
X_7	a,b	a,b,d	—	—	—	c,d

$$(a+bc)\cdot(d+bc)=bc+ad$$

In the Boolean expression. The element of the class (B) is only $\{b,c\}$ in the relation (X_4,X_1). Then, the following equation is derived.

$$b\cdot(bc+ad)=bc+abd$$

and

$$c\cdot(bc+ad)=bc+acd \tag{5.6}$$

Thus, the derived reducts become $\{bc,\ abd,\ acd\}$.

5.2.2 *Modified Reduct Based on Reducts*

In our previous studies [12–14], reducts—nearest neighbor classification is useful. But, its classification does not always show the higher values. To improve the classification accuracy, the reducts are developed to the modified reducts, in which more variables are added to the reducts as follows. The attribute sequence b≻c≻a≻d is obtained, where b≻c implies that the occurrence of attribute b is greater or equal to that of c in the number of its occurrences. That of c is greater than that of a. Similarly, based on c in reduct {b, c}, c ≻ b ≻ d ≻ a is obtained. Further, based on d in reduct {b, d}, d ≻ b ≻ c ≻ a is obtained. Since the c ≻ b ≻ d, and d ≻ b ≻ c consist of the same set of {b, c, d}, the set counted twice. Thus, the modified reduct is adopted as {b, c, d}. The modified reduct {b, c, d} is shown in Fig. 5.3, which is related to reducts, {b, c} and {b, d} [12, 13]. Based on the reducts {b, c} and {b, d}, the following three modified reducts are derived on the discernibility matrix.

$$\{\mathrm{b, c, a}\},\ \{\mathrm{b, c, d}\}\ \text{and}\ \{\mathrm{b, d, a}\} \tag{5.7}$$

The classification accuracy is improved by using modified reducts. The accuracy of the reducts {b, d} and {b, c} becomes 0.56 and 0.78, respectively, while that of the modified reducts {b, c, a} and {b, c, d} becomes 0.95 and 0.86, respectively.

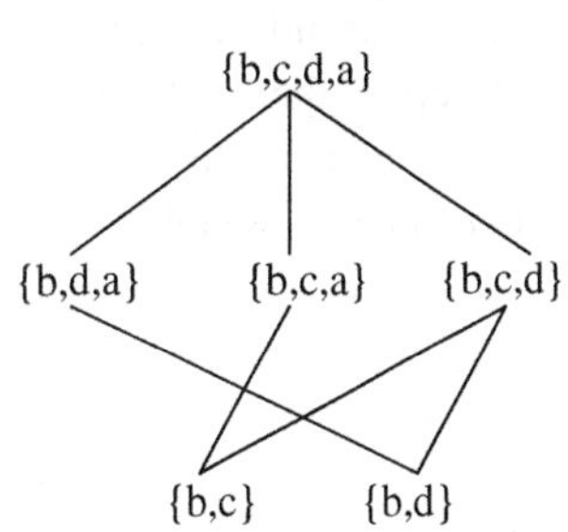

Fig. 5.3 Relation among reducts, modified reducts and the extended reduct

5.3 Linearly Separable Condition in Data Vector Space

To improve the classification accuracy, the data vector space is developed here, in which the classification is characterized by the data independent vectors. Then, the basic analysis is based on the linear separable condition in the data vector space.

Two data sets A and B are said to be linearly separable if their convex hulls are disjoint, that is if $C(A) \cap C(B) = \phi$, where $C(Z)$ shows a convex hull of the set Z and ϕ is a null. The following theorem is mentioned in our study of pattern recognition with threshold elements [15].

Theorem 5.3 *Let S be a subset of the set* Ω *of all the inputs* $\{X_n\}$, *then the linearly separable condition holds for all possible dichotomies of the set S if and only if the vectors* $\{\underline{X_n}\}$ *are linearly independent, where the vector* $\underline{X_n}$ *is defined as the ones that emanate from an arbitrary vertex and terminate at the other inputs,* $X_n \in S$, $X_n \neq X_n^0$.

This is proved by rewriting the separable condition $C(A) \cap C(B) = \phi$ into the following equations. Let Ω be the set of inputs X_n and divide Ω into two sets, A and B. The element of the convex hull $C(A)$ of the set A, is described in,

$$\sum_{X_n^l \in A} C_l X_n^l$$

where C_j is a scalar value. Similarly, the element of the convex hull $C(B)$ of the set B, is represented by

$$\sum_{X_n^m \in B} C_m X_n^m$$

Then, the linearly separable condition is described as follows: If the equations

$$\sum_{X_n^l \in A} C_l X_n^l = \sum_{X_n^m \in B} C_m X_n^m$$

and

$$\sum C_l = \sum C_m, \quad C_l \geq 0, \quad C_m \geq 0 \tag{5.8}$$

hold if and only if all the scalars, C_l and C_m, are zero.

Suppose that set S consists of (n + 1) inputs in U_n, $X_n^0, X_n^1, X_n^2, \ldots, X_n^n$. Then, the vectors emanate from an arbitrary input denoted here by X_n^0 and terminate at the other inputs denoted here by $X_n^1, X_n^2, \ldots, X_n^n$, are given by $\underline{X}^i = X_n^i - X_n^0$, $i = 1, 2, \ldots, n$.

If the vectors $\underline{X}^i$ are linearly independent, the equation holds if and only if $C_i = 0,\ i = 1, 2, \ldots, n$.

$$\sum_{i=1}^{n} C_i \underline{X}^i = 0 \tag{5.9}$$

We can assume X_n^0 to be origin after the linear transformation without losing generality. Then, Eq. (5.2) becomes

$$\sum_{i=1}^{n} C_i X^i = 0, i = 1, 2, \ldots, n \tag{5.10}$$

This equation is divided to terms of A and B inputs subsets. Thus, since all the coefficients C_i are zero, the linear independence satisfies the linearly separable condition, which is a sufficient condition. The necessary condition is shown in the following. If the linear independence is not satisfied, the following equation holds.

$$\sum_{i=1}^{n} C_i X^i = 0, i = 1, 2, \ldots, n \tag{5.11}$$

with some $C_i \neq 0$. Then, the equation is described as follows,

$$V = \sum_{X^l \in A} C_l X^l = \sum_{X^m \in B} C_m X^m \quad C_l \geq 0, C_m \geq 0 \tag{5.12}$$

Since all the C_l and C_m are not zero, the intersection V exists between $C(A)$ and $C(B)$ from Eq. (5.12). This does not satisfy the linear separable condition.

5.4 Nonlinear Mapping of Reducts Based on Nearest Neighbor Relation

5.4.1 Generation of Independent Vectors Based on Nearest Neighbor Relation

In the previous section, it is shown that independent data vectors are useful in the classification. To generate independent data vectors, a nonlinear mapping is developed here based on the reduct. Instances with the nearest neighbor relation are needed to be independent vectors. As an example, the reduct {b, c} in the Eq. (5.5) is adopted, which is shown in Table 5.4.

From Table 5.4, lexicographical ordering of reduct {b, c} is shown in Fig. 5.4.

Table 5.4 Data of reduct {b, c}

Attrib / Inst.	b	c	Class
X_1	0	2	+1
X_2	0	2	+1
X_3	2	0	−1
X_4	2	2	−1
X_5	1	0	−1
X_6	1	1	+1
X_7	1	2	−1

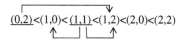

(0,2)<(1,0)<(1,1)<(1,2)<(2,0)<(2,2)

Fig. 5.4 Lexicographical ordering of reduct{b, c} data

In Fig. 5.4, the nearest neighbor relation with minimal distance is generated as shown in directed arrows between +1 class and −1 class. The relation consists of 4 data, (0, 2), (1, 2), (1, 1) and (1, 0). Then, the problem is how to generate the nonlinear variables. The polynomial variables are candidate of the nonlinear mapping. Thus, two variables $\{bc, c^2\}$ is adopted here. Thus, a nonlinear mapping $\{b, c, bc, c^2\}$ is generated from the reduct $\{b, c\}$ as shown. Then, the nonlinear mapping $\{b, c, bc, c^2\}$ is represented in Table 5.5. Then, the nonlinear mapping of the reduct is represented in Table 5.5. In Fig. 5.5, the nonlinear mapping is realized

Table 5.5 Nonlinear mapping of reduct {b, c}

Var. / NNR	b	c	bc	c^2	*class*
(0, 2)	0	2	0	4	+1
(1, 2)	1	2	2	4	−1
(1, 1)	1	1	1	1	+1
(1, 0)	1	0	0	0	−1

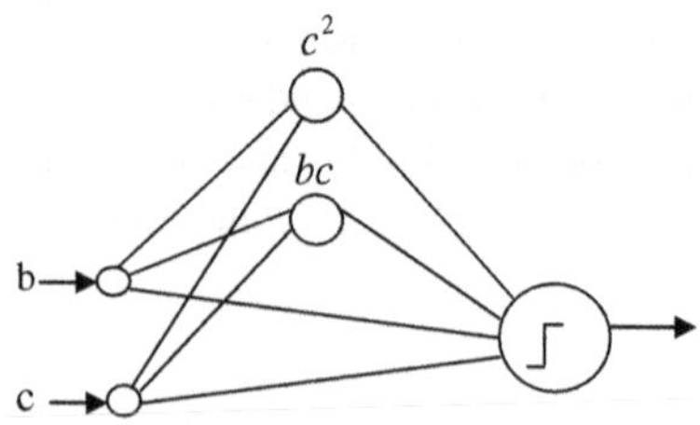

Fig. 5.5 Nonlinear mapping of reduct {b, c}

by a threshold element with the weighted summation of the input {b, c} and nonlinear terms $\{bc, c^2\}$ summation and threshold of function with +1 and −1. The nonlinear mapping of reduct is generalized based on generation of independent input vectors by Theorem 5.3, which is realized by products of variables of reduct. The generalized result is stated in the Theorem 5.4.

Theorem 5.4 *A nonlinear mapping of the reduct is realized by the products of variables of the reduct and a threshold element.*

To make effective clustering after the nonlinear mapping, it will be useful to make the independent vectors. The variable b in Table 5.3 is regarded as the vector $\{b\} = (0111)$. Similarly, the variable $\{c\} = (2210)$, $\{bc\} = (0201)$ and $\{c^2\} = (4410)$ are also regarded as vectors. These vectors become independent vectors, since we can compute the LU decomposition [16] from the matrix of the vectors $\{b, c, bc, c^2\}$. The matrix is full rank(= 4*) in* Table 5.3.

5.4.2 *Characterized Equation of Nearest Neighbor Relation for Classification*

To improve the classification accuracy are assumed to be b and c, since the reduct is {b, c}. The first step is to make the nonlinear data set as shown in Table 5.5, the weighted transformation is introduced. The nearest neighbor relations must satisfy the following inequalities.

$$\begin{aligned} 0w_1 + 2w_2 + 0w_3 + 4w_4 &\geq +1 \\ 1w_1 + 2w_2 + 2w_3 + 4w_4 &\leq -1 \\ 1w_1 + 1w_2 + 1w_3 + 1w_4 &\geq +1 \\ 1w_1 + 0w_2 + 0w_3 + 0w_4 &\leq -1 \end{aligned} \tag{5.13}$$

where the value at the right side in each inequality, +1 or −1 indicate + class or −1 class, respectively. In case of the equality of Eq. (5.13), the weights $\{w_1, w_2, w_3, w_4\}$ of the hyperplane are derived as

$$\{w_1 = -1, w_2 = 9/2, w_3 = -1/2, w_4 = -2\} \tag{5.14}$$

Thus, the hyperplane is generated by the weight values of Eq. (5.14) and threshold θ, which dichotomize sets of the nearest neighbor relation. To extend to all the data classification, the characterized equation is introduced by using nonlinear terms. The variable $\{x_i\}$ of the hyperplane is replaced by $\{b, c, bc, c^2, \ldots\}$, computed $\{w_i\}$ and θ value Then, the characterized equation becomes

$$w_1 b + w_2 c + w_3 bc + w_4 c^2 + \ldots \geq \theta \, (or < \theta) \tag{5.15}$$

All the data except nearest neighbor relation is checked whether Eq. (5.15) classifies correctly. Then the following theorem is derived,

Theorem 5.5 *All the data is classified correctly, if the set with the nearest neighbor relation satisfies independent vector equation as Eq. (5.13) and the characterized Eq. (5.15) is satisfied by all the data except the set of nearest neighbor relation.*

As an example, by replacing $x_1 = b, x_2 = c, x_3 = bc, x_4 = c^2$ and the obtained $\{w_i\}$ in Eq. (5.13) and $\theta = 0$, the characterized equation becomes,

$$-b + \frac{9}{2}c - \frac{1}{2}bc - 2c^2 \geq 0 \, (or < 0) \tag{5.16}$$

which is checked by all the data and classified correctly.

5.4.3 Data Characterization on Nearest Neighbor Relation

To cope with noise in classification, instance data may be adjusted or pruned. This adjust processing is carried out based on the nearest neighbor relation vectors. Other data except nearest neighbor data is characterized by the set of data of the nearest neighbor relation, which forms the independent vectors in the matrix in Table 5.3. For example, a testing data (b, c) = (2, 3) becomes a vector data, $(2\,3\,6\,9)^t$ by the nonlinear mapping. Then, the vector data is represented with independent vectors $(0\,2\,0\,4)^t$, $(1\,2\,2\,4)^t$, $(1\,1\,1\,1)^t$ and $(1\,0\,0\,0)^t$, which are row components made of the nearest neighbor relation in Table 5.5. Since the matrix in Table 5.5 is full in rank, these vectors become independent. Thus, the testing vector data, $(2\,3\,6\,9)^t$ is represented as follows,

$$\begin{pmatrix} 2 \\ 3 \\ 6 \\ 9 \end{pmatrix} = m_1 \overset{+1 \text{ class}}{\begin{pmatrix} 0 \\ 2 \\ 0 \\ 4 \end{pmatrix}} + m_2 \overset{-1 \text{ class}}{\begin{pmatrix} 1 \\ 2 \\ 2 \\ 4 \end{pmatrix}} + m_3 \overset{+1 \text{ class}}{\begin{pmatrix} 1 \\ 1 \\ 1 \\ 1 \end{pmatrix}} + m_4 \overset{-1 \text{ class}}{\begin{pmatrix} 1 \\ 0 \\ 0 \\ 0 \end{pmatrix}} \tag{5.17}$$

where m_1, m_2, m_3 and m_4 are coefficients of independent vectors. Then, the coefficients become

$$m_1 = -\frac{3}{2}, m_2 = -\frac{9}{2}, m_3 = -3 \quad \text{and} \quad m_4 = \frac{1}{2} \tag{5.18}$$

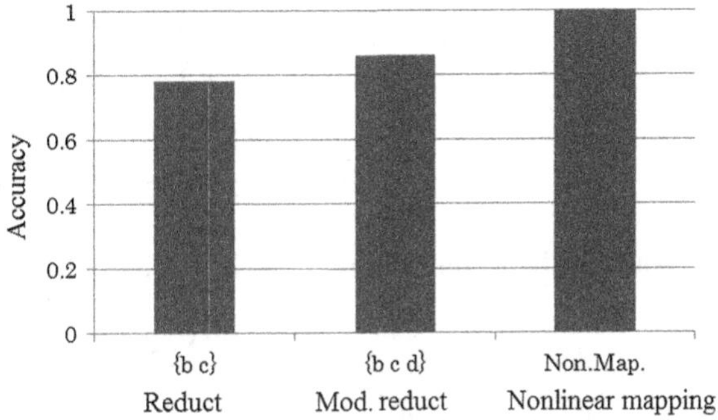

Fig. 5.6 Comparison of classification accuracy among reduct, modified reduct and nonlinear mapping

Since the vector $(0\,2\,0\,4)^t$ belongs to +1 class, the vector is assumed to be in the + directive vector. But, since the coefficient $m_1(-3/2)$ is minus, the vector $m_1(0\,2\,0\,4)^t$ becomes a—directive vector. Next, since the vector $(1\,2\,2\,4)^t$ belongs to the −1 class, the vector shows a—directive vector. Since the coefficient $m_2(9/2)$ is plus, the vector $m_2(1\,2\,2\,4)^t$ belongs to the—directive vector. Similarly, the vector $m_3(1\,1\,1\,1)^t$ belongs to the—directive vector and the vector $m_4(1\,0\,0\,0)^t$ belongs to the—directive vector. Thus, the testing vector composes from these four minus directive vectors, which can be considered to be a minus deep vector in the −1 class. This is clarified that the weight $W \cdot X = (w_1\,w_2\,w_3\,w_4)(2\,3\,6\,9)^t$ in Eq. (5.14), becomes $-(19/2)$, which shows far from minus vectors computations $W \cdot X = -1$ of the nearest neighbor relation and the weighted values are derived from the nearest neighbor relation vectors. $m_1(0\,2\,0\,4)^t$, $m_2(1\,2\,2\,4)^t$ and $m_4(1\,0\,0\,0)^t$ belong to the—directive vector, while $m_3(1\,1\,1\,1)^t$ belong to a + directive vector. This shows data ordering can be made in the nonlinear space. Classification accuracy based on nearest neighbor relation is compared among reduct, modified reduct and nonlinear mapping based on the reduct developed here. The modified reduct was developed [12, 13] for the improvement of the classification accuracy of the reduct {b, c} in Fig. 5.6. Further, the classification accuracy is improved in the nonlinear mapping.

5.4.4 Making Boundary Margin

Boundary margin is often discussed for classification. We develop an asymmetric boundary margin for the classification developed here. To make a deep minus depth than the value $-\frac{1}{2}$, the following equations are developed. Since $m_1(0\,2\,0\,4)^t$, $m_2(1\,2\,2\,4)^t$ and $m_4(1\,0\,0\,0)^t$ with Eqs. (5.17) and (5.18) contribute to the minus deep value of the testing vector, the variable $K(\geq 1)$ is set for $W = (w_1\,w_2\,w_3\,w_4)$ in the following equation.

$$\begin{aligned}0w_1 + 2w_2 + 0w_3 + 4w_4 &\geq +K\\ 1w_1 + 2w_2 + 2w_3 + 4w_4 &\leq -K\\ 1w_1 + 1w_2 + 1w_3 + 1w_4 &\geq +1\\ 1w_1 + 0w_2 + 0w_3 + 0w_4 &\leq -K\end{aligned} \tag{5.19}$$

And for the testing vector $X' = (2, 1, 2, 1)$, the following equation is needed.

$$W \cdot X' < -K \tag{5.20}$$

From Eq. (5.19) with equality, a solution W is derived as follows,

$$W = \left\{ -K \frac{4+5K}{2} \frac{-K}{2} - (1-K) \right\} \tag{5.21}$$

From Eqs. (5.20) and (5.21), the K value is derived as follows,

$$K > 2$$

Thus, by setting $K = 3$, $W = \left(-3 - \frac{19}{2} - \frac{3}{2} - 4 \right)$ is obtained. Then, for the testing vector $X' = (2, 1, 2, 1)$,

$$W \cdot X' = -3.5 \tag{5.22}$$

which shows a deep minus value than the value, $-K = -3$ of the independent vectors based on nearest neighbor relations in Eq. (5.19). For other testing vectors, $W \cdot (2\,0\,0\,0)^t = -6$, $W \cdot (2\,2\,2\,4)^t = -9$, and $W \cdot (2\,3\,6\,9)^t = -22.5$ hold. This shows data ordering preserves in the nonlinear space. Classification accuracy based on nearest neighbor relation is compared among reduct, modified reduct and nonlinear mapping based on the reduct developed here.

5.5 Nonlinear Embedding of Reducts and Threshold Element

A simplified nonlinear mapping of the reduct is realized by conjunction with other variables. As the nonlinear mapping of the reduct, the nonlinear operation of the products $\{c^2\}$ is shown in the dotted box in Fig. 5.7, which we call here an embedding of the nonlinear variables. The nonlinear variable c^2 is made from the variable of the reduct {b, c} in Table 5.4 as shown in Fig. 5.7. Classification accuracies of reduct {b, c}, modified reduct {b, c, d} in Sect. 5.2.2 and nonlinear mapping in Fig. 5.7 are compared as shown in Fig. 5.6. Nonlinear mapping improves the classification accuracy. The summation of the nonlinear variable c^2 and other variables {b, c, a, d} is realized in the final threshold element. The output of this network is followed by the nearest neighbor classifier. The classification of reduct data $\{X_1, X_2, X_3, \ldots, X_6, X_7\}$ with variables $\{c^2$, b, c, a, d$\}$ is carried out by weighting each variable. Then, the network in Fig. 5.7 is followed by the nearest neighbor processing for +1 or −1 class for the respective instances. The final classification accuracy in Fig. 5.8 followed by the nearest neighbor classification is 1.0. This shows the embedding of the nonlinear variable c^2 plays an important role for the classification.

To compare the accuracy of the classification with and without nonlinear mapping, the classification of the data $\{X_1, X_2, X_3, \ldots, X_6, X_7\}$ with variables {b, c, a, d} followed by the nearest neighbor is carried out in Fig. 5.7, in which the dotted square of the nonlinear mapping is removed. Then, the classification accuracy is 0.71 as shown in Fig. 5.8, when only variables {b, c, a, d} of the instances in Table 5.1 are used for the nearest neighbor classification, while the embedding c^2

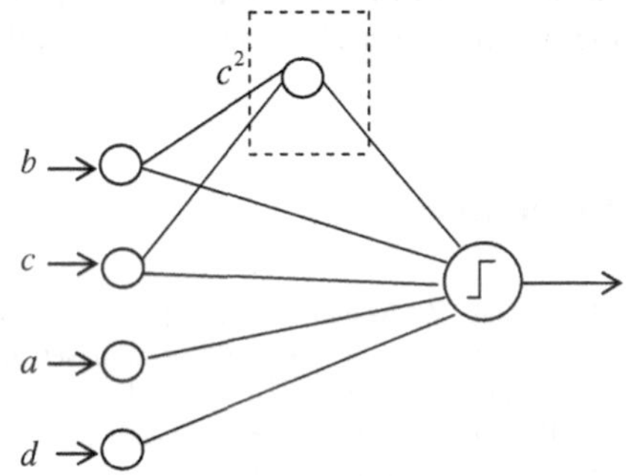

Fig. 5.7 Nonlinear embedding of reduct with other variables and threshold element

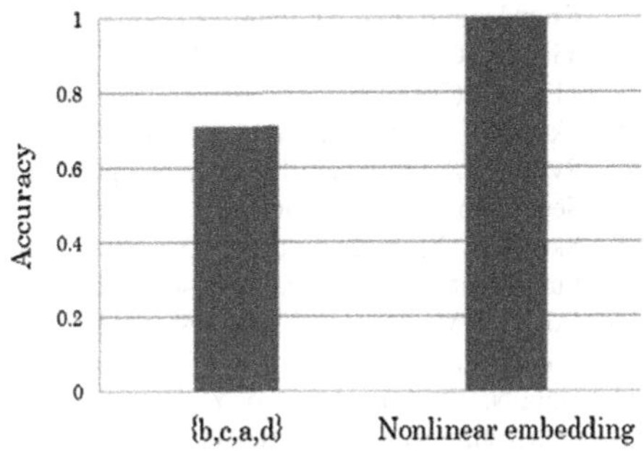

Fig. 5.8 Classification accuracy by variables {b, c, a, d} and nonlinear embedding {b, c} with other variables {a, d}

with other variables {b, c, a, d} shows 1.0 in the accuracy as shown in Fig. 5.8. This shows the embedding of the nonlinear variables from the reduct is effective for classification.

5.6 Conclusion

Reduct is introduced as the minimal set for the data classification in the rough set theory. For the reduction of data dimension, the reduct is useful for the classification. For the data classification, reducts followed the nearest neighbor classification is simple and effective for the processing. In this paper, nearest neighbor relation with minimal distance is proposed, which is the set of data instances with minimal distance in the different classes. Reducts are generated effectively based on the nearest neighbor relation. To improve the classification accuracy of the reduct followed nearest neighbor classification, nonlinear mapping of variables of the reduct is developed, which is based on the construction of independent data vector of the reduct. To analyze the mapping easily and effectively, vector approaches are carried out in the nonlinear transformation. Finally, it is shown that embedding of the nonlinear variables also improves the classification accuracy.

References

1. Pawlak, Z.: Rough Sets. Int. J. Comput. Inf. Sci. **11**, 341–356 (1982)
2. Pawlak, Z., Slowinski, R.: Rough set approach to multi-attribute decision analysis. Eur. J. Oper. Res. **72**, 443–459 (1994)
3. Pawlak, Z.: Some issues on rough sets. In: Transactions on Rough Sets I, LNCS, vol. 3100, pp. 1–58. Springer (2004)
4. Pawlak, Z.: Rogh Sets. Theoretical Aspects of Reasoning About Data. Kluwer Academic Publishers, Boston (1991)
5. Yao, Y., Zhao, Y.: Attribute reduction in decision-theoretic rough set models. Inf. Sci. **178**, 3356–3373 (2008)
6. Jia, X., Shang, L., Zhou, B., Yao, Y.: Generalized attribute reduct in rough set theory. In: Knowledge-Based Systems, vol. 91, pp. 204–218 (2016)
7. Yin, Y., Gong, G., Han, L.: Control approach to rough set reduction. Comput. Math. **57**, 117–127 (2009)
8. Skowron, A., C. Rauszer, C.: The Discernibility Matrices and Functions in Information Systems, in Intelligent Decision Support- Handbook of Application and Advances of Rough Sets Theory, pp. 331–362. Kluwer Academic Publishers, Dordrecht (1992)
9. Skowron, A., Polkowski, L.: Decision algorithms, a survey of rough set theoretic methods. fundamenta. Informatica **30**(3–4), 345–358 (1997)
10. Cover, T.M., Hart, P.E.: Nearest neighbor pattern classification. IEEE Trans. Inf. Theory **13** (1), 21–27 (1967)
11. Ishii, N., Morioka, Y., Bao, Y., Tanaka, H.: Control of variables in reducts-kNN classification with confidence. In: KES2011, LNCS, vol. 6884, pp. 98–107. Springer (2011)
12. Ishii, N., Torii, I., Bao, Y., Tanaka, H.: Modified reduct and their processing for nearest neighbor classification. In: IDEAL, LNCS, vol. 7435, pp. 753–762. Springer (2012)

13. Ishii, N., Torii, I., Bao, Y., Tanaka, H.: Mapping of nearest neighbor relation. In: Proceedings of ICIS, pp. 71–76 (2013)
14. Ishii, N., Torii, I., Nakashima T., Iwata, K.: Generation of reducts based on nearest neighbor relation. In: Proceedings of ACIS-IEEE SNPD, pp. 1–6 (2014)
15. Kimura, M., Ishii, N., Honda, N.: Pattern recognition with threshold elements. In: Proceedings of ICMCI, pp. 135–142 (1964)
16. Poole, D.: Linear Algebra. A Modern Introduction, 2nd edn. Canada Thomson Books/Cole (2006)

Chapter 6
New Quality Indexes for Optimal Clustering Model Identification Based on Cross-Domain Approach

Jean-Charles Lamirel

Abstract Feature maximization is an alternative measure, as compared to usual distributional measures relying on entropy or on Chi-square metric or vector-based measures, like Euclidean distance or correlation distance. One of the key advantages of this measure taking inspiration both from Galois lattice theory and information retrieval is that it is operational in an incremental mode on traditional classification. In this framework, it does not have the limitations of the aforementioned measures in the case of the processing of highly unbalanced, heterogeneous and highly multidimensional data. We present a new cross-domain application of this measure in the clustering context for setting up new cluster quality indexes that are tolerant to noise and whose efficiency ranges from low to high dimensional data.

6.1 Introduction

Unsupervised classification or clustering is a data analysis technique which is increasingly widely-used in different areas of application. If the datasets to be analyzed have growing size, it is clearly unfeasible to get ground truth that permits to work on them in a supervised fashion. The main problem which then arises in clustering is to qualify the obtained results in terms of quality. A quality index is a criterion which makes possible to decide which clustering method to use, to fix an optimal number of clusters and also to evaluate or develop a new method. Many approaches have been developed for that purpose as it has been pointed out in [1, 23, 24, 27]. However, even if recent alternative approaches do exist [4, 12, 13], the usual quality indexes are mostly based on the concepts of dispersion of a cluster and dissimilarity between clusters. Computation of the latter criteria themselves relies on Euclidean distance. Most popular such indexes are the Dunn index [8], the Davis-Bouldin index [6], the Silhouette index [25], the Calinski-Harabasz index [5] and the Xie-Beni index [28]. They implement the aforementioned concepts in slightly different ways.

J.-C. Lamirel (✉)
SYNALP Team, LORIA, Bâtiment B, 54506 Vandœuvre-lès-Nancy, France
e-mail: lamirel@loria.fr

I. Hatzilygeroudis et al. (eds.), *Advances in Combining Intelligent Methods*,
Intelligent Systems Reference Library 116, DOI 10.1007/978-3-319-46200-4_6

The Dunn index (DU) identifies clusters which are well separated and compact. It combines dissimilarity between clusters and their diameters to estimate the most reliable number of clusters. The Davies-Bouldin index (DB) is similar to the Dunn index and identifies clusters which are far from each other and compact. The Silhouette index (SI) computes a width depending on the membership of a data point in any cluster. A negative silhouette value for a given point means that the point is most suited to belong to a different cluster from the one it is allocated. The Calinski-Harabasz index (CH) computes a weighted ratio between the within-group scatter and the between group scatter. Well separated and compact clusters should maximize this ratio. The Xie-Beni index (XI) is a compromise between the approaches provided by the Dunn index and by the Calinski-Harabasz index.

As stated in [11, 27] usual indexes have the defect to be sensitive to noisy data and outliers. In [19], we also observed that the proposed indexes are not suitable to analyze clustering results in highly multidimensional space as well since they are unable to detect degenerated clustering results. Also these indexes are not independent of the clustering method with which they are used. As an example, a clustering method which tends to optimize WGSS, like k-means [22], will also tend to naturally produce low value for that criteria which optimizes indexes output, but does not necessarily guarantee coherent results, as it was also demonstrated in [19]. Last but not least, as Hamerly et al. pointed out in [14], the experiments on these indexes in the literature are often performed on unrealistic test corpora made up of low dimensional data with a small number of "well-shaped" (mostly hyperspheric) embedded virtual clusters. As an example, in their reference paper, Milligan and Cooper [23] compared 30 different methods for estimating the number of clusters. They classified CH and DB in the top 10, with CH the best but their experiments only used simulated data described in a low dimensional Euclidean space. The same remark can be made about the comparison performed in [27] or in [7]. However, Kassab et al. [15] used the Reuters test collection to show that the aforementioned indexes are often unable to identify an optimal clustering model whenever the dataset is constituted by complex data which need to be represented in both high-dimensional and sparse description space, obviously with embedded non-Gaussian clusters, as is often the case with textual data. The silhouette index is considered one of the more reliable indexes among those mentioned above especially in the case of multidimensional data, mainly because it is not a diameter-based index optimized for Gaussian context. However, like the Dunn and Xie-Beni indexes, its main defect is that it is computationally expensive, which could represent a major drawback for use with large datasets constituted by high-dimensional data.

There are also other alternatives to the usual indexes. For example, in 2009 Lago-Fernãndez et al. [17] proposed a method using negentropy which evaluates the gap between the cluster entropy and entropy of the normal distribution with the same covariance matrix, but again their experiments were only conducted on two-dimensional data. Also other recent indexes attempts were limited by the researchers' choice of complex parameters [27].

Our aim was to get rid of the method-index dependency problem and the issue of sensitivity to noise while also avoiding computation complexity, parameter settings and dealing with a high-dimensional context. To achieve goals, we exploited features of the data points attached to clusters instead of information carried by cluster centroids and replaced Euclidean distance with a more reliable quality estimator based on the feature maximization measure. This measure has been already successfully used by Lamirel et al. to solve complex high-dimensional classification problems with highly imbalanced and noisy data gathered in similar classes thanks to its very efficient feature selection and data resampling capabilities [21]. As a complement to this information, we shall show in the upcoming experimental section that cluster quality indexes relying on this measure do not possess any of the defects of usual approaches including computational complexity.

Section 6.2 presents a feature maximization measure and our proposed new indexes. Section 6.3 presents our first experimental context based on reference datasets. Section 6.4 details our first results. Section 6.5 draws our conclusion and ideas for future work.

6.2 Feature Maximization for Feature Selection

Feature maximization is an unbiased measure which can be used to estimate the quality of a classification whether it be supervised or unsupervised. In unsupervised classification (i.e. clustering), this measure exploits the properties (i.e. the features) of data points that can be attached to their nearest cluster after analysis without prior examination of the generated cluster profiles, like centroids. Its principal advantage is thus to be totally independent of the clustering method and of its operating mode.

Consider a partition C which results from a clustering method applied to a dataset D represented by a group of features F. The feature maximization measure favours clusters with a maximal feature F-measure. The feature F-measure $FF_c(f)$ of a feature f associated with a cluster c is defined as the harmonic mean of the feature recall $FR_c(f)$ and of the feature predominance $FP_c(f)$, which are themselves defined as follows:

$$FR_c(f) = \frac{\Sigma_{d \in c} W_d^f}{\Sigma_{c \in C} \Sigma_{d \in c} W_d^f} \qquad FP_c(f) = \frac{\Sigma_{d \in c} W_d^f}{\Sigma_{f' \in F_c, d \in c} W_d^{f'}} \tag{6.1}$$

with

$$FF_c(f) = 2\left(\frac{FR_c(f) \times FP_c(f)}{FR_c(f) + FP_c(f)}\right) \tag{6.2}$$

where W_d^f represents the weight of the feature f for the data point d and F_c represents all the features present in the dataset associated with the cluster c.

There is some important similarities between Recall and Predominance used in the proposed approach and Recall and Precision used in information retrieval. We have already exploited this analogy more thoroughly in some of our former works, like in [18], but the measures proposed here must be considered as generalizations of such information retrieval measures which are no more based on agreement but on influence of a feature materialized by a weight. Weight represents the importance of a feature for a data and furthermore for a cluster. The choice of the weighting scheme is not really constrained by the approach instead of producing positive values. Such scheme is supposed to figure out the significance (i.e. semantic and importance) of the feature for the data.

Feature recall is a scale independent measure but feature predominance is not. We have however shown experimentally in [21] that the F-measure which is a combination of these two measures is only weakly influenced by feature scaling. Nevertheless, to guaranty full scale independent behavior for this measure, data must be standardized.

Feature maximization measure can be exploited to generate a powerfull feature selection process [21]. In the clustering context, this kind of selection process can be defined as non-parametrized process based on the content of clusters in which a cluster feature is characterized using both its capacity to discriminate between clusters ($FP_c(f)$ index) and its ability to faithfully represent the cluster data ($FR_c(f)$ index). The set S_c of features that are characteristic of a given cluster c belonging to a partition C is translated by:

$$S_c = \left\{ f \in F_c \mid FF_c(f) > \overline{FF}(f) \text{ and } FF_c(f) > \overline{FF}_D \right\} \tag{6.3}$$

where

$$\overline{FF}(f) = \Sigma_{c' \in C} \frac{FF_{c'}(f)}{|C_{/f}|} \text{ and } \overline{FF}_D = \Sigma_{f \in F} \frac{\overline{FF}(f)}{|F|} \tag{6.4}$$

where $C_{/f}$ represents the subset of C in which the feature f occurs.

Finally, the set of all selected features S_C is the subset of F defined by:

$$S_C = \cup_{c \in C} S_c. \tag{6.5}$$

In other words, the features judged relevant for a given cluster are those whose representations are better than average in this cluster, and better than the average representation of all the features in the partition, in terms of feature F-measure. Features which never satisfy the second condition in any cluster were discarded.

A specific concept of contrast $G_c(f)$ can be defined to calculate the performance of a retained feature f for a given cluster c. It is an indicator value which is proportional

Shoes_ Size	Hair_ Length	Nose_ Size	Class
9	5	5	M
9	10	5	M
9	20	6	M
5	15	5	W
6	25	6	W
5	25	5	W

FR(S,M) = 27/43 = 0.65

FP(S,M) = 27/78 = 0.35

$$FF(S,M) = \frac{2(FR(S,M) \times FP(S,M))}{FR(S,M) + FP(S,M)} = 0.45$$

Fig. 6.1 Principle of feature F-measure computation for sample data

to the ratio between the F-measure $FF_c(f)$ of a feature in the cluster c and the average F-measure $\overline{FF}$ of this feature for the whole partition.[1] It can be expressed as:

$$G_c(f) = FF_c(f)/\overline{FF}(f) \tag{6.6}$$

The active features of a cluster are those for which the contrast is greater than 1. Moreover, the higher the contrast of a feature for one cluster, the better its performance in describing the cluster content.

Below we give an example of the operating mode of the method, on the basis of a toy-dataset encompassing two classes (*Men (M)*, *Women (F)*) described with 3 features: *Nose_Size*, *Hair_Length*, *Shoes_Size*. Figure 6.1 shows the source data and how the F-measure calculation of the *Shoes_Size* feature operates in the *Men* class.

As shown in Fig. 6.2, the second step in the process is to calculate the marginal average F-measure for each feature and the overall average F-measure for the combination of all features and all classes. In this figure, notation $\overline{F(.,.)}$ stands here for overall average $\overline{FF}_D$ presented in (Eq. 6.3) and notation $\overline{F(x,.)}$ stands for marginal average of class x, which is itself computed as:

$$\overline{F(x,.)} = \Sigma_{f \in S_x} \frac{FF_x(f)}{|S_x|} \tag{6.7}$$

Features with F-measures that are systematically lower than the overall average are eliminated. The *Nose_Size* feature is thus removed. Remaining features (i.e. selected features) are considered active in the classes in which their F-measure is above the marginal average:

---

[1] Using p-value highlighting the significance of a feature for a cluster by comparing its contrast to unity contrast would be a potential alternative to the proposed approach. However, this method would introduce unexpected Gaussian smoothing in the process.

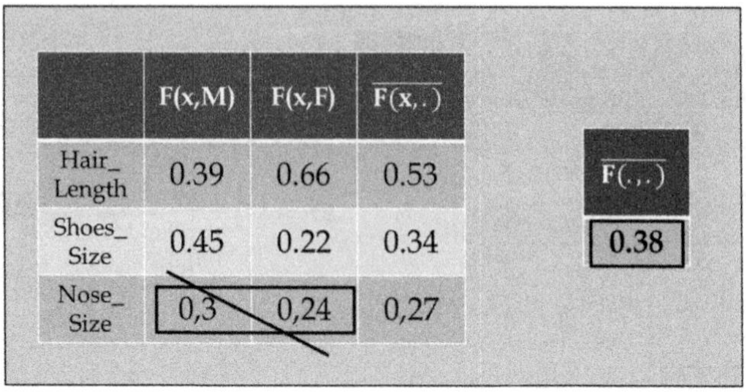

Fig. 6.2 Principle of computation of overall feature F-measure average and elimination of irrelevant features

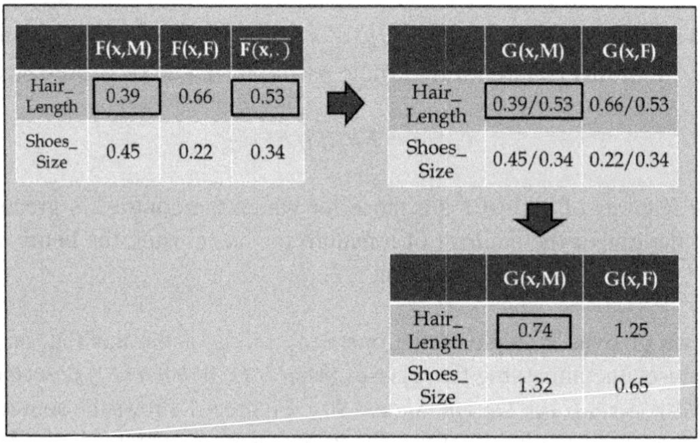

Fig. 6.3 Principle of computation of contrast for selected features

1. *Shoes_Size* is active in the *Men's* class,
2. *Hair_Length* is active in the *Women's* class.

Contrast ratio highlights the degree of activity and passivity of selected features as regards their F-measure marginal average in different classes. Figure 6.3 illustrates how the contrast is calculated for the example presented. In the context of this example, the contrast may be considered as a function that will virtually have the following effects:

1. Increase the length of women's hair,
2. Increase the size of men's shoes,
3. Decrease the length of men's hair,
4. Reduce the size of women's shoes.

As already mentioned before, the active features in a cluster are selected features for which the contrast is greater than 1 in that cluster. Conversely, the passive features in a cluster are selected features present in the cluster's data for which contrast is less than unity.[2] A simple way to exploit the features obtained is to use active selected features and their associated contrast for cluster labelling as we proposed in [21]. A more sophisticated method (as we shall propose hereafter) is to exploit information related to the activity and passivity of selected features in clusters to define clustering quality indexes identifying an optimal partition. This kind of partition is expected to maximize the contrast described by Eq. 6.6. This approach leads to the definition of two different indexes: PC and EC index.

The PC index, whose principle corresponds by analogy to that of intra-cluster inertia in the usual models, is a macro-measure based on the maximization of the average weighted contrast of active features for optimal partition. For a partition comprising k clusters, it can be expressed as:

$$PC_k = \frac{1}{k} \sum_{i=1}^{k} \frac{1}{n_i} \sum_{f \in S_i} G_i(f) \tag{6.8}$$

The EC index, whose principle corresponds by analogy to that of the combination between intra-cluster inertia and inter-cluster inertia in the usual models, is based on the maximization of the average weighted compromise between the contrast of active features and the inverted contrast of passive features for optimal partition:

$$EC_k = \frac{1}{k} \sum_{i=1}^{k} \left(\frac{\frac{|s_i|}{n_i} \sum_{f \in S_i} G_i(f) + \frac{|\overline{s_i}|}{n_i} \sum_{h \in S_i} \frac{1}{G_i(h)}}{|s_i| + |\overline{s_i}|} \right) \tag{6.9}$$

where n_i is the number of data associated with the cluster i, $|s_i|$ represents the number of active features in i, and $|\overline{s_i}|$, the number of passive features in the same cluster.

6.3 Experimental Data and Process

To objectively calculate the accuracy of our new indexes, we used several different datasets of varying dimensionality and size for which the optimal number of clusters (i.e. ground truth) is known in advance.

A part of the datasets came from the UCI machine learning repository [3] and is more usually exploited for classification tasks. The four selected UCI datasets represent mostly low to middle dimensional datasets and small datasets (except for PEN dataset which is large). The ZOO and SOY datasets which include variables with modalities are transformed into binary files. IRIS is exploited both in standard

---

[2] As regards the principle of the method, this type of selected features inevitably have a contrast greater than 1 in some other cluster(s) (see Eq. 6.3 for details).

Table 6.1 Datasets overall characteristics (Binarization of IRIS dataset results in 12 binary features out of 4 real-valued features)

	IRIS	IRIS-b	WINE	PEN	SOY	ZOO	VRBF	R8	R52
Nbr class	3	3	3	10	16	7	12–16	8	52
Nbr data	150	150	178	10992	292	101	2183	7674	9100
Nbr feat	4	12	13	16	84	114	231	3497	7369

and in binarized version to obtain clearer insight into the behavior of quality index on binary data.

The VERBF dataset is a dataset of French verbs which are described both by semantic features and by subcategorization frames. The ground truth of this dataset has been established both by linguists who studied different clustering results and by a gold standard based on the VerbNet classification, as in [26]. This binary dataset contains verbs described in a space of 231 Boolean features. It can be considered a typical middle size and middle dimensional dataset.

The R8 and R52 corpora were obtained by Cardoso Cachopo from the R10 and R90 datasets, which are derived from the Reuters 21578 collection.[3] The aim of these adjustments was to only retain data that had a single label. Considering only monothematic documents and classes that still had at least one example of training and one of test, R8 is a reduction of the R10 corpus (the 10 most frequent classes) to 8 classes and R52 is a reduction of the R90 corpus (90 classes) to 52 classes. The R8 and R52 are large and multidimensional datasets with respective size of 7674 and 9100 and associated bag of words description spaces of 1187 and 2618 words. These datasets can be considered large and high dimensional.

The summary of the overall characteristics of datasets is provided in Table 6.1.

We exploited two different usual clustering methods, namely k-means [22], a winner-take-all method, and GNG [10], a winner-take-most method with Hebbian learning. For text and/or binary datasets we also used the IGNGF neural clustering method [19] which has already been proven to outperform other clustering methods, including spectral methods [26], on this kind of data. We have reported on the method that produced the best results in the following experiments.

As class labels were provided in all datasets and considering that the clustering method could only produce approximate results as compared to reference categorization, we also used purity measures to estimate the quality of the partition generated by the method as regards to category ground truth. Following [26], we use modified purity (mPUR) to evaluate the clustering results produced and this was computed as follows:

$$mPUR = \frac{|P|}{|D|} \tag{6.10}$$

[3]http://www.research.att.com/~lewis/reuters21578.html.

where $P = \{d \in D \mid \text{prec}(c(d)) = g(d) \wedge |c(d)| > 1\}$ with D being the set of exploited data points, $c(d)$ a function that provides the cluster associated to data point d and $g(d)$ a function that provides the gold class associated to data point d. Clusters for which the prevalent class has only one element are considered as marginal and are thus ignored.

For the same reason, we also varied the number of clusters in a range up to three times that determined by the ground truth. An index which gave no indication of optimum in the expected range was considered to be out-of-range or diverging index (- out-). We finally obtained a process which consists of generating disturbance in the clustering results by randomly exchanging data between clusters to different fixed extents (10, 20, 30 %) whilst maintaining the original size of the clusters. This process simulated increasingly noisy clustering results and the aim was to estimate the robustness of the proposed estimators.

6.4 Results

The results are presented in Tables 6.2, 6.3 and 6.4. Some complementary information is required regarding the validation process. In the tables, MaxP represents the number of clusters of the partition with highest mPUR value (Eq. 6.10), or in some cases, the interval of partition sizes with highest stable mPUR value. When a quality index identified an optimal model with MaxP clusters and MaxP differed from the number of categories established by ground truth, its estimation was still considered valid. This approach took into account the fact that clustering would quite systematically produce sub-optimal results as compared to ground truth. The partitions with the highest purity values were thus studied to deal with this kind of situation. For a similar reason, all estimations in the interval range between the op-

Table 6.2 Overview of the indexes estimation results on low dimensional data (Bold numbers represent valid estimations)

	IRIS	IRIS-b	WINE	PEN	SOY	Number of correct matches
DB	2	5	**5**	7	**19**	**2/5**
CH	2	**3**	6	8	5	**1/5**
DU	1	1	8	17	8	**0/5**
SI	**4**	2	7	14	14	**1/5**
XB	2	7	-out-	19	24	**0/5**
PC	**3**	**3**	**4**	9	**16**	**4/5**
EC	**3**	**3**	**4**	9	**16**	**4/5**
MaxP	3	3	5	11	19	
Method	K-means	K-means	GNG	GNG	GNG	

Table 6.3 Overview of the indexes estimation results on average to high dimensional data (Bold numbers represent valid estimations)

	ZOO	VRBF	R8	R52	Number of correct matches
DB	**8**	-out-	5	58	**1/4**
CH	4	7	**6**	-out-	**1/4**
DU	**8**	2	-out-	-out-	**1/4**
SI	4	-out-	-out-	**54**	**1/4**
XB	-out-	23	-out-	-out-	**0/4**
PC	7	18	-out-	-out-	**1/4**
EC	**7**	**15**	**6**	**52**	**4/4**
MaxP	10	12–16	6	50–55	
Method	IGNGF	IGNGF	IGNGF	IGNGF	

Table 6.4 Indexes estimation results in the presence of noise (UCI ZOO dataset)

	ZOO	ZOO Noise 10 %	ZOO Noise 20 %	ZOO Noise 30 %	Number of correct matches
DB	**8**	4	3	3	**1/4**
CH	4	5	3	3	**0/4**
DU	**8**	2	2	2	**1/4**
SI	14	-out-	-out-	-out-	**0/4**
XB	-out-	-out-	-out-	-out-	**0/4**
PC	6	4	11	**9**	**1/4**
EC	**7**	5	6	**9**	**2/4**
MaxP	10	7	10	10	
Method	IGNGF	IGNGF	IGNGF	IGNGF	

timal k (ground truth) and MaxP values were also considered valid. When indexes were still increasing and decreasing (depending on whether they were maximizers or minimizers) when the number of clusters was more than three times the number of expected classes, they were considered out-of-range (-out- symbol in Tables 6.2, 6.3 and 6.4). The Fig. 6.4 depicts the trends of evolution of EC and PC indexes in the case of the R52 dataset. It highlights what is a suitable index behaviour (EC index) and in a parallel way what represents the out-of-range index behaviour we mentioned before (PC index).

When considering the results presented in Tables 6.2 and 6.3, it should first be noted that one of our tested indexes, the Xie-Beni (XB) index never provides any correct answers. These were either out of range (i.e. diverging) or answers (i.e. minimum value when this index was a minimizer) in the range of the variation of k, but too far from ground truth or even too far from optimal purity among the set of generated clustering models. Some indexes were in the low mid-range of correctness and provide unstable answers. This was the case with the Davis-Bouldin (DB),

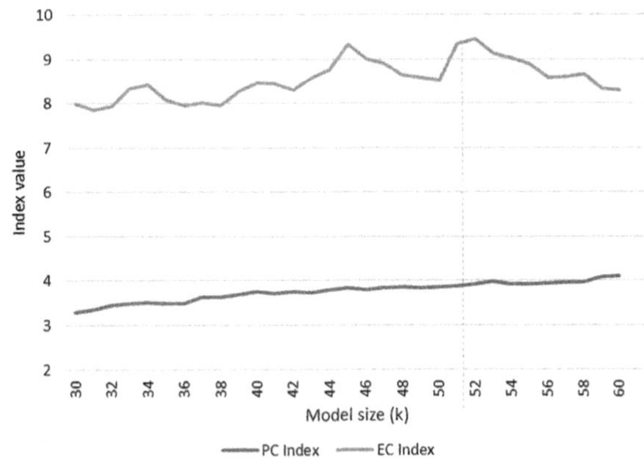

Fig. 6.4 Trends of PC and EC indexes on Reuters R52 dataset

Calinski-Harabasz (CH), Dunn (DU) and Silhouette (SI) indexes. When dimension increased, these indexes were found to become generally unable to provide any correct estimation. This phenomenon has already been observed in previous experiments with Davis-Bouldin (DB) and Calinski-Harabasz (CH) indexes [15]. Davis-Bouldin (DB) performed slightly better than average on low dimensional data. Our PC index was found to perform significantly better than average on low dimensional data but obviously remains a better low dimensional problem estimator than a high dimensional one. Help from passive features somehow seems mandatory to estimate an optimal model in the case of high dimensional problems. Hence, the EC index which exploited both active and passive features was found to have from far the best performance, whatever it faced with low (Table 6.2) or high dimensional estimation problem (Table 6.3). According to our evaluation criteria, this index only do wrong in the case of the PEN dataset. However, even in this case its estimation (model of size 9) is still in the close neighbour of the optimal one (model of size 10). Additionally, the EC and PC indexes were both found to be capable of dealing with binarized data in a transparent manner which is not the case of some of the usual indexes namely the Xie-Beni (XI) index, and to a lesser extent, Calinski-Harabasz (CH) and Silhouette (SI) indexes.

Interestingly, on the UCI ZOO dataset, the results of noise sensitivity analysis presented in Table 6.4 underline the fact that noise has a relatively limited effect on the operation of PC and EC indexes. The EC index was again found to have the most stable behavior in that context. The Fig. 6.5 presents a parallel view of the different trends of EC value on non noisy and noisy clustering environment, respectively. It shows that noise tends to lower the index value in an overall way and to soften the trends related to its behaviour relatively to changes in k value. However, the index is still able to estimate, either the optimal model in the best case, or a neighbour model in the worst case. The usual indexes do not work as well at all in the same context. For example, the Silhouette index firstly delivered the wrong optimal k values on this dataset before getting out of range when the noise reached 20 % on clustering results. The Davis-Bouldin (DB) and Dunn (DU) indexes were found to shift from a correct to a wrong estimation as soon as noise began to appear.

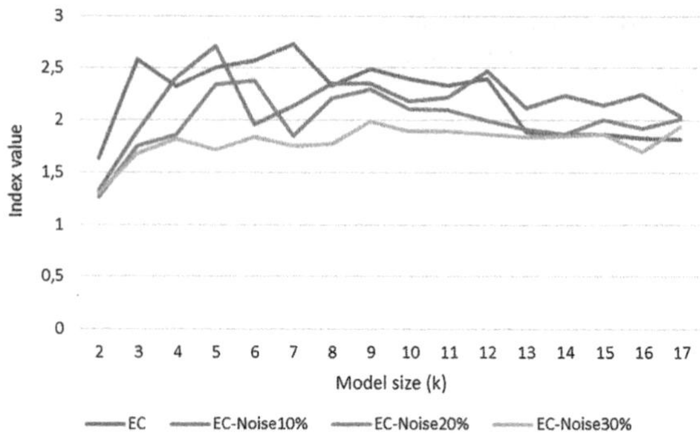

Fig. 6.5 Trends of EC indexes on UCI ZOO dataset with and without noise

In all our experiments, we observed that the quality estimation depends little on the clustering method. Moreover, we noted that the computation time of the index was one of the lowest among the indexes studied. As an example, for the R52 dataset, the EC index computation time was 125 s as compared to 43000 s for the Silhouette index using a standard laptop with 2.2 GHz quadricore processor and 8 GB of memory.

6.5 Conclusion

Using a cross-domain approach taking inspiration both from Galois lattice theory and from information retrieval we have proposed a new set of indexes for clustering quality evaluation relying on feature maximization measurement. This method exploits the information derived from features which could be associated to clusters by means of their associated data. Our experiments showed that most of the usual quality estimators do not produce satisfactory results in a realistic data context and that they are additionally sensitive to noise and perform poorly with high dimensional data. Unlike the usual quality estimators, one of the main advantages of our proposed indexes is that they produce stable results in cases ranging from a low dimensional to high dimensional context and also require low computation time while easily dealing with binarized data. Their stable operating mode with clustering methods which could produce both different and imperfect results also constitutes an essential advantage. However, further experiments are required using both an extended set of clustering methods and a larger panel of high dimensional datasets to confirm this promising behavior.

References

1. Angel Latha Mary, S., Sivagami, A.N., Usha Rani, M.: Cluster validity measures dynamic clustering algorithms. ARPN J. Eng. Appl. Sci. **10**(9) (2015)
2. Arellano-Verdejo, J., Guzmn-Arenas, A., Godoy-Calderon, S., Barrn Fernãndez, R.: Efficiently finding the optimum number of clusters in a dataset with a new hybrid cellular evolutionary algorithm. Computacin y Sistemas, **18**(2):313–327 (2014)
3. Bache, K., Lichman, M.: UCI Machine learning repository. University of California, School of Information and Computer Science, Irvine, CA, USA (2013). http://archive.ics.uci.edu/ml
4. Bock H-H (1996) Probability model and hypothese testing in partitionning cluster analysis. In: Arabie P, Hubert LJ, De Soete G (eds) Clustering and Classification. World Scientific, Singapore, pp 377–453
5. Calinsky T, Harabasz J (1974) A dendrite method for cluster analysis. Commun. Stat. 3(1):1–27
6. Davies, D.L., Bouldin, D.W.: A cluster separation measure. IEEE Trans. Pattern Anal. Mach. Intell. PAMI-1, **2**:224–227 (1979)
7. Dimitriadou E, Dolnicar S, Weingessel A (2002) An examination of indexes for determining the number of clusters in binary data sets. Psychometrika 67(1):137–159
8. Dunn J (1974) Well separated clusters and optimal fuzzy partitions. J. Cybern. 4:95–104
9. Falk I, Lamirel J-C, Gardent C (2012) Classifying French Verbs Using French and English Lexical Resources. Proccedings of ACL, Jeju Island, Korea
10. Fritzke, B.: A growing neural gas network learns topologies. In: Tesauro, G., Touretzky, D.S., Leen, T.K. (eds.) Advances in Neural Information Processing Systems **7**, pp. 625–632 (1995)
11. Guerra L, Robles V, Bielza C, Larrañaga P (2012) A comparison of clustering quality indices using outliers and noise. Intell. Data Anal. 16:703–715
12. Gordon, A.D.: External validation in cluster analysis. Bull. Int. Stat. Inst. **51**(2), 353–356 (1997). (Response to comments. Bull. Int. Stat. Inst. **51**(3), 414–415 (1998))
13. Halkidi M, Batistakis Y, Vazirgiannis M (2001) On clustering validation techniques. J. Intell. Inf. Syst. 17(2/3):147–155
14. Hamerly, G., Elkan, C.: Learning the K in K-Means. In: Neural Information Processing Systems (2003)
15. Kassab, R., Lamirel, J.-C.: Feature based cluster validation for high dimensional data. In: IASTED International Conference on Artificial Intelligence and Applications (AIA), pp. 97–103. Innsbruck, Austria, (February 2008)
16. Kolesnikov A, Trichina E, Kauranne T (2015) Estimating the number of clusters in a numerical data set via quantization error modeling. Pattern Recognit. 48(3):941–952
17. Lago-Fernãndez LF, Corbacho F (2009) Using the negentropy increment to determine the number of clusters. In: Cabestany J, Sandoval F, Prieto A, Corchado JM (eds) Bio-Inspired Systems: Computational and Ambient Intelligence. Springer, Berlin, pp 448–455
18. Lamirel, J.-C., Francois, C., Shehabi Al, S., Hoffmann, M.: New classification quality estimators for analysis of documentary information: application to patent analysis and web mapping. Scientometrics **60**(3), 445–462 (2004)
19. Lamirel, J.-C., Mall, R., Cuxac, P., Safi, G.: Variations to incremental growing neural gas algorithm based on label maximization. in: Proceedings of IJCNN 2011, p. 956–965. San Jose, CA, USA (2011)
20. Lamirel J-C (2012) A new approach for automatizing the analysis of research topics dynamics: application to optoelectronics research. Scientometrics 93(1):151–166
21. Lamirel J-C, Cuxac P, Chivukula AS, Hajlaoui K (2014) Optimizing text classification through efficient feature selection based on quality metric. J. Intell. Inf. Syst. 2013:1–18 Special issue on PAKDD-QIMIE
22. MacQueen, J.B.: Some methods for classification and analysis of multivariate observations. In: Proceedings of 5th Berkeley Symposium on Mathematical Statistics and Probability (1), pp. 281–297. University of California Press (1967)

23. Milligan GW, Cooper MC (1985) An examination of procedures for determining the number of clusters in a dataset. Psychometrika 50(2):159–179
24. Rendón E, Abundez I, Arizmendi A, Quiroz EM (2011) Internal versus External cluster validation indexes. Int. J. Comput. Commun. 5(1):27–34
25. Rousseeuw PJ (1987) Silhouettes: a graphical aid to the interpretation and validation of cluster analysis. J. Comput. Appl. Math. 20:53–65
26. Sun, L., Korhonen, A., Poibeau, T., Messiant, C.: Investigating the cross-linguistic potential of VerbNet-style classification. In: Proceedings of ACL, pp. 1056–1064. Beijing, China (2010)
27. Yanchi, L., Zhongmou, L., Xiong, H., Gao, X., Wu, J.: Understanding of internal clustering validation measures. In: Proceedings of the 2010 IEEE International Conference on Data Mining, ICDM'10, pp. 911–916 (2010)
28. Xie, X.L., Beni, G.: A validity measure for fuzzy clustering. IEEE Trans. Pattern Anal. Mach. Intell. **13**(8), 841–847 (1991)

Chapter 7
A Hybrid User and Item Based Collaborative Filtering Approach by Possibilistic Similarity Fusion

Manel Slokom and Raouia Ayachi

Abstract Collaborative filtering is the most successful recommender system to date. It explores techniques for matching people with similar interests and making recommendations on this basis. Existing collaborative filtering methods are user-based and item-based CF. The user-based calculates prediction based on similarity weight between each pair of users. The item-based finds items rated by the target user that are similar to the item being predicted. However, in both cases only partial information from user-item matrix is employed to predict unknown ratings. Therefore, it seems desirable to fuse preferences from both similar users and similar items. These preferences are generally certain and perform accurate predictions. But this does not reflect the reality which is related to uncertainty and imprecision by nature. Consequently, this paper proposes a novel approach, called Π HCF, that combines on the one hand, user-based and item-based collaborative filtering recommendation performance and on the other hand possibility theory in order to cope with the uncertain aspect of user-item preferences. Experimental results demonstrate that the proposed method gives a better prediction and shows a considerable improvement of recommendation.

Keywords Recommendation systems · Collaborative filtering · User-based · Item-based · Hybrid · Possibility theory · Uncertain preferences

---

M. Slokom (✉) · R. Ayachi
LARODEC, Institut Supérieur de Gestion Tunis, 2000 Le Bardo, Tunisie
e-mail: manel.slokom@live.fr

R. Ayachi
e-mail: raouia.ayachi@gmail.com

I. Hatzilygeroudis et al. (eds.), *Advances in Combining Intelligent Methods*, Intelligent Systems Reference Library 116, DOI 10.1007/978-3-319-46200-4_7

7.1 Introduction

Recommender systems have emerged in the past several years as an efficient tool used to deliver users with more intelligent and proactive information service. Such systems apply knowledge discovery techniques to personalize recommendations provided for each user. Recommender systems use several algorithms to help us sort the masses of information and find the "good product" in a very personalized way [1, 14]. They recommend products or services that fit well the learned users' preferences and needs. These systems generally combine, on the one hand, information extracted from users' profiles and social interactions, and on the other hand, machine learning techniques that are used to predict the user's ratings or preferences.

Several types of recommenders have been proposed in the literature that can be categorized into three major categories [14], namely *content-based filtering*, *collabo rative filtering*, and *hybrid filtering* approaches. In this work, we focus on collaborative filtering as it is considered as the most popular approach [8]. The main idea consists in finding desired items on the basis of the similar users' preferences set. In fact, collaborative filtering methods compare other users' preferences degrees then, select the most likely items and generate the recommendation list. Collaborative filtering algorithms are divided into two main categories, namely *memory*-based and *model*-based. The first category focuses on the entire collection of previously rated items, while the model-based one uses a model learned from the collection of ratings to make predictions. In this work, we are in particular interested in memory-based approaches based on user-item matrix which represents users preferences.

Several memory-based collaborative filtering techniques, either user-based or item-based, exist in the literature [12, 13, 18]. User-based methods exploit historical users, whose behaviors are similar to the one of the target user in order to predict his preferences degrees. While item-based methods make use of similar items to the target item to ensure the prediction. However, these methods are quite often not deterministic, since ratings are generally unavailable as both user-based and item-based CF algorithms only take into consideration a very small portion of ratings. Therefore, predictions are often made from not so similar users or items. Consequently, to solve these problems, recent few researches proposed to combine the two types of predictions into a single value. In fact, authors in [22] presented an adaptive fusion mechanism to capture 3-dimensional correlations between users, items and tags. Also, in [23] authors combined user-based and item-based CF techniques in order to improve recommendation diversity by combining popular items of user and item based collaborative filtering methods. Another aim behind combining user and item -based CF approach is to alleviate the *data sparsity* problem, and this idea was proposed in [10]. Based on this combination, the recommender system is able to propose relevant recommendations for users.

Most of standard collaborative filtering techniques give good results in a certain context presupposing that the available preferences are precisely and exactly defined, which is not always realistic. In fact, when dealing with real-world applications, users' preferences can be imprecise and/or uncertain or even missing. Thus, the recommendation results are deeply affected if uncertainty is not considered. Consequently, a good recommender should be able to suggest items even when information about ratings is imperfect. For these reasons, few studies [16, 24] have introduced uncertainty in the recommendation process. However, to the best of our knowledge, no research studied a purely uncertain hybrid recommender system dealing with uncertain ratings as input.

In this paper, we propose a new approach named "A possibilistic combination of user-based and item-based collaborative filtering recommender", denoted by Π HCF, which uses the *possibility theory* framework to cope with the uncertainty that may pervade users ratings. In fact, our approach is based on four steps, namely:

1. *Preferences representation* handles users preferences using the possibility theory framework.
2. *Possibilistic predictions* are divided into two steps:
   - A possibilistic item-based collaborative filtering build on the assumption that a person will favor the items that are similar to the items he liked in the past.
   - A possibilistic user-based collaborative filtering build on the assumption that each person belongs to a large group sharing similar interests.
3. *Information fusion* is a unified framework incorporating both user and item-based collaborative filtering algorithms.
4. *Recommendation generation* consists in recommending the top-K most likely items that can interest the target user.

The rest of this paper[1] is organized as follows. In Sect. 7.2, we will briefly present background knowledge related to collaborative filtering and the possibility theory framework. Section 7.3 is dedicated to previous hybrid research works. We describe our possibilistic collaborative filtering approach Π HCF in Sect. 7.4. Section 7.5 is dedicated to the experimental study. Finally, we conclude the paper in Sect. 7.6.

7.2 Background Knowledge

This section gives a brief overview on both collaborative filtering approach and the possibility theory.

7.2.1 Collaborative Filtering

Recommender systems (for short RS) are a class of information filtering system looking to predict the ratings that user would give to an item [14]. The goal of RS is to provide the user with a list of recommendations that might meet their preferences,

[1]This is an extended and revised version of the conference paper [19].

or to suggest predictions on how much the user might prefer each item. The most used techniques are [1, 14]: collaborative filtering and content-based systems. In this paper, we are particularly interested to the collaborative filtering approach (CF) as it is the most successful and popular recommendation technique to date [8, 10, 23], which aims at predicting users' preferences on items by exploiting the historical rating information.

In a CF recommender system, preferences data of users towards items are represented as a *user-item rating matrix*. Typical collaborative filtering approaches are [8, 23]:

7.2.1.1 User-Based Collaborative Filtering

Searches the most similar users of the target user, then calculates rating prediction based on similarity between each pair of users. Central aspects to these algorithms are [4]:

- how to identify neighbors forming the best strategy to generate item recommendations for the target user.
- how to make use of the information provided by them.

In what follows, we detail the standard user-based process composed of two main components: neighborhood identification and prediction computation:

- **Neighborhood identification**: is based on choosing the co-rated users who are similar to the target user according to a similarity metric. The similarity S, between two users is generally computed by finding a set of items that both users have interacted with, then examining to what degree the users displayed similar behaviors on these items. Several metrics are proposed to calculate the similarity or distance between users such as *pearson correlation* and *cosine distance*. In this paper, pearson correlation coefficient is adopted, since several previous studies [2, 21, 25] confirm that this latter performs better than others. In fact, such metric measures the extent to which two variables relate with each other [21]. Formally, the Pearson correlation between two users u and u' is:

$$S(u,u') = \frac{\sum_{i\in I'}(r_{u,i}-\overline{r}_u)(r_{u',i}-\overline{r}_{u'})}{\sqrt{\sum_{i\in I'}(r_{u,i}-\overline{r}_u)^2}\sqrt{\sum_{i\in I'}(r_{u',i}-\overline{r}_{u'})^2}} \tag{7.1}$$

  where I' is the set of items that both u and u' have rated, $r_{u,i}$ is the rating of the user u for the item i and $\overline{r}_u$ is the average rating of the co-rated items of the user u.
- **Prediction computation**: To make a prediction for the target user u' on a certain item i, we consider the weighted sum of all other rated items on that item [21] according to the following formula:

$$P_{u',i} = \overline{r}_{u'} + \frac{\sum_{u\in U}(r_{u,i}-\overline{r}_u) * S(u,u')}{\sum_{u\in U}|S(u,u')|} \tag{7.2}$$

Table 7.1 User-item preferences matrix

	i_1	i_2	i_3	i_4	i_5
u_1	$\pi_{u_1}(i_1) = 1$	$\pi_{u_1}(i_2) = 2$	$\pi_{u_1}(i_3) = ?$	$\pi_{u_1}(i_4) = 2$	$\pi_{u_1}(i_5) = 1$
u_2	$\pi_{u_2}(i_1) = 2$	$\pi_{u_2}(i_2) = ?$	$\pi_{u_2}(i_3) = 5$	$\pi_{u_2}(i_4) = 4$	$\pi_{u_2}(i_5) = 3$
u_3	$\pi_{u_3}(i_1) = 3$	$\pi_{u_3}(i_2) = 5$	$\pi_{u_3}(i_3) = 4$	$\pi_{u_3}(i_4) = 3$	$\pi_{u_3}(i_5) = ?$
u_4	$\pi_{u_4}(i_1) = ?$	$\pi_{u_4}(i_2) = 2$	$\pi_{u_4}(i_3) = 5$	$\pi_{u_4}(i_4) = 3$	$\pi_{u_4}(i_5) = ?$

where $\overline{r}_u$ and $\overline{r}_{u'}$ are the average ratings for the user u' and user u on all other rated items.

Example 1 Considering the preference matrix of Table 7.1. We want to compute the u_2's prediction for item i_2, using the user-based collaborative filtering algorithm. To this end, we proceed as follows:

$\widehat{x}_{2,2} = \overline{x}_2 + \frac{s(u_2,u_1)(x_{1,2}-\overline{x}_1)+s(u_2,u_3)(x_{3,2}-\overline{x}_3)+s(u_2,u_4)(x_{4,2}-\overline{x}_4)}{|s(u_2,u_1)+s(u_2,u_3)+s(u_2,u_4)|} = 4$ where $s(u_2, u_1) = \mathbf{0.87}$, $s(u_2, u_3) = -0.29$ *and* $s(u_2, u_4) = -1$.

7.2.1.2 Item-based Collaborative Filtering

Makes predictions for the target user based on how he has rated items that are similar to the target item [18]. The item similarity computation is defined in terms of ratings correlations between the target item and the other items. Similarities between items are computed using the previously evoked metrics. Here, we will just present the prediction computation component.

- **Prediction computation**: The simple weighted average is used to predict the rating $P_{u,i}$ for user u on item i. Formally:

$$P_{u,i} = \frac{\sum_{j \in N} r_{u,j} * S(i,j)}{\sum_{j \in N} |S(i,j)|} \tag{7.3}$$

where the summations are over all other rated items $j \in N$ for user u, $S(i,j)$ is the similarity between items i and j, $r_{u,j}$ is the rating for user u on item j.

Example 2 Considering the same example of Table 7.1, we want to compute $\widehat{\pi}_{u_2}(i_2)$ prediction based on item-based CF algorithm: $\widehat{x}_{2,2} = \frac{s(i_2,i_1)*x_{2,1}+s(i_2,i_3)*x_{2,3}+s(i_2,i_5)*x_{2,5}}{|s(i_2,i_1)+s(i_2,i_3)+s(i_2,i_5)|} = \mathbf{3}$ where $s(i_2, i_1) = 0$, $s(i_2, i_3) = -0.8$ and $s(i_2, i_5) = 1$.

7.2.2 Possibility Theory

Originally, managing uncertainty has especially been explored in artificial intelligence. Depending on the nature of imperfection and uncertainty, several tools have been proposed such as, probability theory, fuzzy set theory [17], evidence theory [20] and possibility theory [7]. During the recent years, these well established theories have started to play a key role in modeling and treating uncertainty in reasoning processes. Among the aforementioned uncertainty theories, we are interested in *possibility theory* which offers a natural and simple tool to handle uncertain information. It differs from the probability theory framework by the use of a pair of dual set-functions (possibility and necessity measures) instead of only one. Indeed, in possibility theory, experts can express their uncertainty numerically using possibi lity degrees or qualitatively using orderings on the possible values. These define the two interpretations of possibility theory, namely, the *quantitative* and *qualitative* settings. In this paper, we use the *quantitative possibility theory*. In this section, we provide basic concepts relative to the possibility theory (for more details see [7]). Let Ω denotes the universe of discourse, which is the Cartesian product of all variable domains in V. Each element $\omega \in \Omega$ is called a state of Ω. $\omega[X_i] = x_i$ denotes an instantiation of X_i in ω. We denote ϕ, α the sub classes of Ω called events and $\neg\phi$ denotes the complementary set of ϕ i.e., $\neg\phi = \Omega - \phi$.

7.2.2.1 Possibility Distribution

The basic building block in the possibility theory is the concept of *possibility distribution* π, which corresponds to a function associating to each element ω_i from the universe of discourse Ω a value to a bounded and linearly ordered valuation set (L, <).

Contrary to the standard probability theory, the possibilistic scale could be interpreted in twofold: a *numerical interpretation* when values have a real sense (L = [0, 1]) and an *ordinal* one ($>_\pi$) when values only reflect a total pre-order between the different states of the world. In this work, we will focus on the numerical interpretation.

The degree $\pi(\omega)$ represents the compatibility of ω with available pieces of information. By convention, $\pi(\omega) = 1$ means that ω is totally possible and $\pi(\omega) = 0$ means that ω is an impossible state. If $\pi(\omega) > \pi(\omega')$, this means that ω is preferred to ω'. In the possibility theory framework, there are two extreme cases:

- *Complete knowledge*: $\exists\omega_0$, $\pi(\omega_0) = 1$ and $\pi(\omega) = 0$ $\forall\omega \neq \omega_0$ (only ω_0 is possible).
- *Total ignorance*: $\forall\omega \in \Omega$, $\pi(\omega) = 1$ (all states are possible).

A possibility distribution π is said to be *normalized* if there exists at least one totally possible state. Formally:

$$\exists\omega \in \Omega, \pi(\omega) = 1 \tag{7.4}$$

Example 3 $\pi(T1 = win) = 1, \pi(T1 = equalize) = 0.7, \pi(T1 = lose) = 0.2, \pi(T1 = win) = 1$ means that it is fully possible for team 1 to win the game. The possibility distribution given by the arbiter is normalized since max $(1, 0.7, 0.2) = 1$.

7.2.2.2 Inconsistency

In possibility theory, the inconsistency is measured by the degree of conflict between uncertain information.

$$Inc(\pi) = 1 - max_{\omega \in \Omega}\, \pi(\omega) \tag{7.5}$$

In this case, π is considered as *sub-normalized*, otherwise, π is said to be *normalized* (i.e. $max_{\omega \in \Omega}\, \pi(\omega) = \pi(\omega_i) = 1$). It is clear that, for normalized π, $max_{\omega \in \Omega}\, \pi(\omega) = 1$, hence $Inc(\pi) = 0$. The measure *Inc* is very useful in computing the conflict between two distributions π_1 and π_2 given by $Inc(\pi_1, \pi_2) = Inc(\pi_1 \wedge \pi_2)$, where $\wedge$ is a conjunctive t-norm operator. For simplicity, we take the *minimum* conjunctive ($\wedge$) operator. Obviously, when $\pi_1 \wedge \pi_2$ gives a sub-normalized possibility distribution, it indicates that there is a conflict between π_1 and π_2. On the other hand, $\pi_1 \wedge \pi_2$ is normalized, there is no conflict and hence $Inc(\pi_1, \pi_2) = 0$.

Example 4 Let $\pi_1[1, 0.2, 0.5]$ and $\pi_2[0.8, 0, 0.3]$ be two possibility distributions. We take the minimum as the conjunctive operator, we obtain: $Inc(\pi_1, \pi_2) = Inc([0.8, 0, 0.3]) = 1 - 0.8 = 0.2$. Thus, the two sources are inconsistent with each other.

7.2.2.3 Information Fusion

Is used to combine possibility distributions issued from distinct sources and which should pertain to the same variable. There are different combination modes proposed to deal with the problem of possibilistic information fusion. The choice of the combination mode is related to the assumption about the reliability of sources [6].

In possibility theory, the two basic combination modes are conjunctive and disjunctive, each of which has some specific merging operators. In general, *conjunctive operators* (e.g. minimum, product) are advised to combine information that is reliable, consistent and agree with each other [5, 15]. On the other hand, the *disjunctive operators* (e.g. maximum, probabilistic sum, etc.) are applied when it is believed for sure that at least one of the sources is reliable but it is not known which one and when there is a high degree of conflict among sources. Thus, disjunctive operators are advised to merge inconsistent information. In fact, the degree of inconsistency of merged information is widely used to assess how consistent that two pieces of information are. Obviously, this value is not sufficient when multiple pairs of uncertain information have the same degree of inconsistency [15]. Consequently, authors in [5] elaborated an *adaptive operator* which allows a switch from conjunctive fusion to disjunctive fusion, according to the amount of conflict $Inc(\pi_1 \wedge \pi_2)$ existing between the two sources S_1 and S_2, when neither of them is suitable for merging alone.

Table 7.2 The conjunctive, disjunctive and adaptive fusion

	π_1	π_2	$\pi_\vee$	$\pi_\wedge$	π_{AD}
$T1 = win$	1	0.8	$max(1; 0.8) = 1$	$\frac{min(1;0.8)}{0.8} = 1$	$max(1; 0.2) = \mathbf{1}$
$T1 = equalize$	0.7	1	$max(0.7; 1) = 1$	$\frac{min(0.7;1)}{0.8} = 0.875$	$max(0.875; 0.2) = \mathbf{0.875}$
$T1 = lose$	0.2	0.5	$max(0.2; 0.5) = 0.5$	$\frac{min(0.2;0.5)}{0.8} = 0.25$	$max(0.25; 0.2) = \mathbf{0.25}$

Therefore, the adaptive fusion operator is the following:

$$\forall \omega \in \Omega; \pi_{AD}(\omega) = max(\pi_\wedge(\omega); min(\pi_\vee(\omega), 1 - h(\pi_1, \pi_2))) \tag{7.6}$$

where:

- $\pi_\wedge(\omega) = \frac{min(\pi_1(\omega), \pi_2(\omega))}{h(\pi_1, \pi_2)}$ is the normalized conjunctive fusion where:
  - $min(\pi_1(\omega), \pi_2(\omega))$ is the conjunctive fusion operator
  - $h(\pi_1, \pi_2) = 1 - Inc(\pi_1, \pi_2)$ expresses the degree of agreement of the two sources
- $\pi_\vee(\omega) = max(\pi_1(\omega), \pi_2(\omega))$ is the disjunctive fusion mode

Example 5 Let us continue with Example 3. We suppose that another arbiter S_2 provide his opinion regarding the game result for team 1, in the form of a possibility distribution, denoted by $\pi_2(T1)$. Thus, we obtain the two possibility distributions π_1 and π_2.

In the Table 7.2, we first calculate the conjunctive fusion between π_1 and π_2 in column 4. Then in column 5 we compute the disjunctive fusion in order to lead to the adaptive fusion in column 6.

7.3 Related Work

Collaborative filtering [8, 21] has attracted a considerable amount of researches resulting in a large variety of collaborative filtering approaches. These methods are often not deterministic. In fact, given an unknown target rating to be estimated, they first measure similarities between target user and other users (user-based), or, between target item and other items (item-based). Then, they compute prediction by averaging ratings from not-so-similar users or not-so-similar items. In both cases, only a very small portion information from the user-item matrix are taken into consideration to predict unknown ratings. To overcome these problems, researchers have recently attempted to combine user- and item-based CF approaches. In fact, Hu and Lu [10] proposed the hybrid predictive user and item based CF with smoothing sparse data (HSPA) algorithm, which computes unknown user's preferences based on combining the strength of both user-based and item-based CF algorithm. In fact, this framework aims to provide the data smoothing using the item-based methods, then predicts the model based on both users' aspects and items' aspects in order to ensure robust predictive accuracy. In the same context, Tso-Sutter et al. [22] described a new method of personalized recommendation using users, items and tags where tags are local descriptions of items given by users. Thus, they proposed at first a generic method allowing tags to be incorporated to standard CF algorithms. Then, they proposed an adaptive fusion mechanism to capture the 3-dimensional correlations between users, items and tags. Also Wang and Yin [23] combined popular items of both item and user -based CF methods in order to propose a more diverse recommendation list of items for target user. By item popularity, it means the users' rating frequency for each item in the system. Yamashita et al. [25] presented an adaptive fusion method for user-based and item-based collaborative filtering in order to improve the recommendation accuracy. The method uses a weight parameter to unify both algorithms. They first investigated the relationship between recommendation accuracy and weight parameter. Then, they proposed an estimation of the appropriate weight value based on absolute collected ratings.

These recent hybrid methods give good results in a certain context, presupposing that the available ratings, from which prediction will be induced, are precisely and exactly defined, which is not always realistic. In fact, when dealing with real-world applications, users' preferences are inseparably connected with imperfection. Thus, ratings can be imprecise and/or uncertain. As a result, the recommendation results are deeply affected. For these reasons, new few studies have considered uncertainty in order to improve the recommender's accuracy. In fact, Wang et al. [24] provided a memory based CF in a probabilistic framework by fusing all absolute ratings from three different sources:

- Prediction based on ratings of the same item by other users
- Prediction based on different item ratings made by the same user
- Ratings predicted based on data from other but similar users ratings and other but similar items

In addition, they introduced two parameters in order to adjust the significance of the three predictions. Last but not least, Miyahara and Pazzani [16] proposed an approach that combines the item-based collaborative filter and the user-based collaborative filter in order to improve the performance of predictions. This approach consists in filling out a pseudo score using both user and item-based collaborative filters using the simple Bayes. Then, the proposed Bayesian collaborative filtering recommender returns the combined results.

These uncertain researches handle the uncertainty aspect only in the fusion process. To the best of our knowledge, no research studied the uncertainty phenomena in users' ratings from the beginning of the hybrid recommendation process. To this end, we propose a new hybrid item-based and user-based recommender system under a possibilistic framework.

7.4 New Possibilistic Combination of User-Based and Item-Based Collaborative Filtering Recommender

Our goal in this work is to take into account the uncertainty that may pervade users' preferences. To assure this task, uncertain preferences will be presented using the possibility theory framework, then a purely possibilistic information fusion mode should be used to merge information issued from several possibilistic preferences. Finally, the unknown preference degree is predicted. Accordingly, we propose a new possibilitic combination of user-based and item-based collaborative filtering approach, denoted Π HCF and based on four phases, namely *preferences representation*, *possibilistic predictions*, *information fusion* and *recommendation generation*.

The first phase consists in building the user-item matrix under a possibilistic framework. The second phase is mainly divided into two steps: *a possibilistic item-based collaborative filtering*, denoted by Π ICF and *a possibilistic user-based collaborative filtering*, denoted by Π UCF. The Π ICF is build on the assumption that a person will favor the items that are similar to the items he liked in the past [24]. The Π UCF is build on the assumption that each person belongs to a large group sharing similar interests [24]. The third phase is a unified framework incorporating both user and item-based CF. The last phase, consists in recommending the top-K most likely items that can interest the target user. The whole process of the proposed Π HCF method is illustrated by the diagram of Fig. 7.1. In this section, we first present some annotation, then we describe the different phases of Π HCF.

7.4.1 Annotation

- M users: $U = \{u_1, .., u_m, .., u_M\}$
- N items: $I = \{i_1, .., i_n, .., i_N\}$
- P: The possibilistic user-item matrix
- $\pi_{u_m}(i_n)$ the preference degree of item i_n provided by user u_m
- $\hat{\pi}^{ICF}_{u_m}(i_n)$: The item-based prediction for item i_n provided by user u_m
- $\hat{\pi}^{UCF}_{u_m}(i_n)$: The user-based prediction for item i_n provided by user u_m
- u^T_m the user u_m profile. It represents user's items preference degrees
- $\pi_{u_m}(i_n) = ?$ means the rating is unknown

7.4.2 Preferences Representation

The representation of users preferences is a primordial step in our approach. In fact, each user should provide his preferences about items as a possibility distribution where each degree corresponds to a satisfaction degree for an item i. When only one

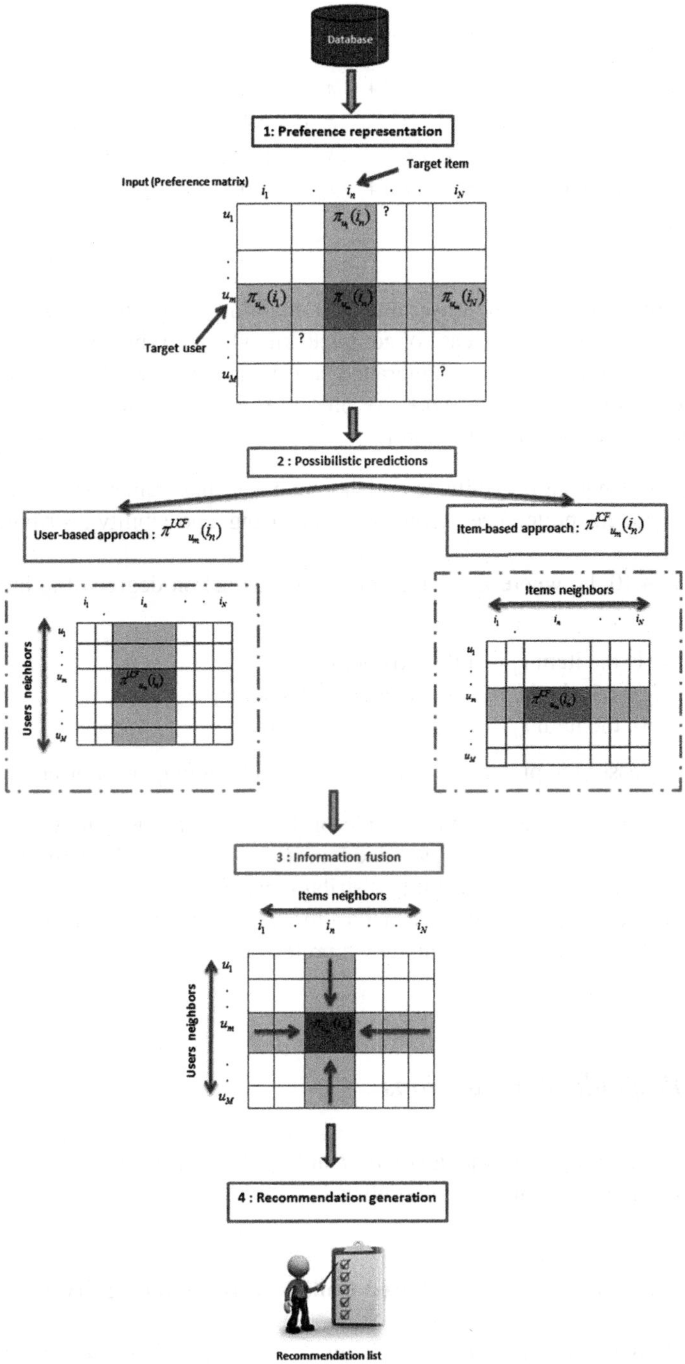

Fig. 7.1 The possibilistic hybrid user-based and item-based collaborative filtering process: Π HCF

Table 7.3 User-item preferences matrix

	i_1	i_2	i_3	i_4	i_5
u_1	$\pi_{u_1}(i_1) = 1$	$\pi_{u_1}(i_2) = 0.4$	$\pi_{u_1}(i_3) = 0.8$	$\pi_{u_1}(i_4) = 0.4$	$\pi_{u_1}(i_5) = 0.2$
u_2	$\pi_{u_2}(i_1) = 0.4$	$\pi_{u_2}(i_2) = ?$	$\pi_{u_2}(i_3) = 1$	$\pi_{u_2}(i_4) = ?$	$\pi_{u_2}(i_5) = ?$
u_3	$\pi_{u_3}(i_1) = 0.6$	$\pi_{u_3}(i_2) = 1$	$\pi_{u_3}(i_3) = ?$	$\pi_{u_3}(i_4) = 0.6$	$\pi_{u_3}(i_5) = 0.7$
u_4	$\pi_{u_4}(i_1) = ?$	$\pi_{u_4}(i_2) = 0.4$	$\pi_{u_4}(i_3) = 0.5$	$\pi_{u_4}(i_4) = 0.6$	$\pi_{u_4}(i_5) = 1$

item is fully satisfactory by a user and all remaining items are not satisfactory at all, we deal with the extreme case of complete knowledge. While when all items are satisfactory by a user, the total ignorance case is tackled. Based on the simplicity of the possibility theory framework, we present users preferences using possibility distributions described in Definition 1.

Definition 1 Under a possibilistic collaborative filtering framework, each user u_m should provide his preferences about an item i_n using a possibility distribution. Formally:

$\pi_{u_m} : I \rightarrow [0, 1]$, where $\pi_{u_m}(i_n)$ denotes the satisfaction degree of i_n by u_m such that:

- $\pi_{u_m}(i_n) = 1$: the item i_n is fully satisfactory.
- $0 < \pi_{u_m}(i_n) < 1$, the item i_n is somewhat satisfactory.
- $\pi_{u_m}(i_n) = 0$, the item i_n is not satisfactory at all.

$\pi(i_n)$ expresses the preference degrees assigned to item i_n by all users.

Example 6 Table 7.3 depicts an example of a possibilistic user-item matrix composed of five users $\{u_1, u_2, u_3, u_4, u_5\}$ and four items $\{i_1, i_2, i_3, i_4\}$, representing users preferences of items using possibility distributions.

For instance, item i_1 is fully satisfactory for user u_4 and not satisfactory at all for user u_1, while $?\, for\ u_2$ means an unknown preference degree.

7.4.3 *Possibilistic Predictions*

In this section, we describe the *item-based* and *the user-based* collaborative filtering processes from a point of view possibilistic.

7.4.3.1 The Possibilistic Item-Based Collaborative Filtering Process

Denoted Π ICF, looks into the set of items the target user has rated and computes how similar they are to the target item by taking into consideration the uncertain aspect of user's preferences.

The critical step in Π ICF approach is to compute the similarity between items. In fact, to compute similarity between items i_1 and i_2 we should first isolate the users who have both rated these items and then apply a possibilistic similarity measure. In our work, we used the possibilistic similarity measure *Information Affinity* as it combines two important criteria, namely distance and inconsistency. This combination is justified by the fact that using only a distance measure does not always decide which is the closest distribution. Intuitively, information affinity takes into account the classical informative distance, e.g. Manhattan or Euclidean which evaluates the difference between two normalized possibility distributions and the inconsistency measure which evaluates the conflict between the possibility distributions. Formally:

$$Aff(\pi(i_n), \pi(i_{n'})) = 1 - \frac{\kappa * d(\pi(i_n), \pi(i_{n'})) + \lambda * Inc(\pi(i_n), \pi(i_{n'}))}{\kappa + \lambda} \quad (7.7)$$

where $\kappa > 0$ and $\lambda > 0$. d, denotes normalized metric distances between $\pi(i_n)$ and $\pi(i_{n'})$. $Inc(\pi(i_n) \wedge \pi(i_{n'}))$ denotes the degree of conflict between the two preference degrees where $\wedge$ is taken as the product or min conjunctive operators.

The unknown preference of a target item by a target user can be predicted by averaging the preferences of other similar items rated by this target user. Thus, we use the *Weighted Sum* technique to obtain prediction as expressed in equation (7.8):

$$\widehat{\pi}^{ICF}_{u_m}(i_n) = \frac{\sum_{Allsimilaritems} Aff(\pi(i_n), \pi(i_{n'})) * \pi_{u_m}(i_{n'})}{\sum_{Allsimilaritems} |Aff(\pi(i_n), \pi(i_{n'}))|} \quad (7.8)$$

Example 7 Considering the preference matrix of Table 7.3. We want to compute the user u_2 prediction for item i_2, i_4 and i_5 based on item-based CF using the Information Affinity similarity measure. We obtain:

$$\widehat{\pi}^{ICF}_{u_2}(i_2) = \frac{Aff(\pi_{i_2},\pi_{i_1})\pi_{u_2}(i_1)+Aff(\pi_{i_2},\pi_{i_3})\pi_{u_2}(i_3)+Aff(\pi_{i_2},\pi_{i_5})\pi_{u_2}(i_5)}{|Aff(\pi_{i_2},\pi_{i_1})+Aff(\pi_{i_2},\pi_{i_3})+Aff(\pi_{i_2},\pi_{i_5})|}$$

$$\Rightarrow \widehat{\pi}^{ICF}_{u_2}(i_2) = \frac{1.172}{1.77} = \mathbf{0.706} \text{ where:}$$

- $Aff(\pi_{i_2}, \pi_{i_1}) = 1 - \frac{d(\pi_{i_2},\pi_{i_1})+Inc(\pi_{i_2},\pi_{i_1})}{2} = 1 - \frac{0.5+0.4}{2} = \mathbf{0.55}$ where:
  - $d(\pi_{i_2}, \pi_{i_1}) = \frac{1}{2}(0.6 + 0.4) = 0.5$
  - $Inc(\pi_{i_2}, \pi_{i_1}) = 1 - max\{0.4; 0.6\} = 0.4$
- $Aff(\pi_{i_2}, \pi_{i_3}) = 1 - \frac{d(\pi_{i_2},\pi_{i_3})+Inc(\pi_{i_2},\pi_{i_3})}{2} = 1 - \frac{0.6+0.3}{2} = \mathbf{0.575}$ where:
  - $d(\pi_{i_2}, \pi_{i_3}) = \frac{1}{2}(0.4 + 0.1) = 0.25$
  - $Inc(\pi_{i_2}, \pi_{i_3}) = 1 - max\{0.4; 0.4\} = 0.6$

In the same manner, we compute the others item-based predictions and we obtain: $\widehat{\pi}^{ICF}_{u_2}(i_4) = \mathbf{0.681}$ *and* $\widehat{\pi}^{ICF}_{u_2}(i_5) = \mathbf{0.671}$.

7.4.3.2 The Possibilistic User-Based Collaborative Filtering Process

Denoted Π UCF, predicts a target user's interest for a target item based on similar users profiles and uncertain users preferences. Thus, Π UCF approach computes the similarity between users u_1 and u_2 by isolating the items which have been both rated by these users and then applying *Information Affinity*. Finally, we use the *Weighted Sum of others' ratings* technique to obtain prediction as expressed in equation (7.9):

$$\widehat{\pi}_{u_m}^{UCF}(i_n) = \overline{\pi}(u_m) + \frac{\sum_{Allsimilarusers} Aff(\pi(u_m), \pi(u_{m'})) * (\pi_{u_m}(i_{n'}) - \overline{\pi}(u_{m'}))}{\sum_{Allsimilarusers} |Aff(\pi(u_m), \pi(u_{m'}))|} \tag{7.9}$$

Example 8 Considering the same example of Table 7.3, we want to compute respectively $\widehat{\pi}_{u_2}(i_2)$, $\widehat{\pi}_{u_2}(i_4)$ and $\widehat{\pi}_{u_2}(i_5)$ predictions based on user-based CF algorithm:

$$\widehat{\pi}_{u_2}^{UCF}(i_2) = 0.7 + \frac{Aff(\pi_{u_2}, \pi_{u_1})(\pi_{u_1}(i_2) - 0.56) + Aff(\pi_{u_2}, \pi_{u_3})(\pi_{u_3}(i_2) - 0.725)}{|Aff(\pi_{u_2}, \pi_{u_1}) + Aff(\pi_{u_2}, \pi_{u_3})|}$$

$$\frac{+Aff(\pi_{u_2}, \pi_{u_4})(\pi_{u_4}(i_2) - 0.625)}{| + Aff(\pi_{u_2}, \pi_{u_4})|} = 0.7 + \frac{-0.112 + 0.165 - 0.112}{2} = \mathbf{0.667}$$

where $Aff(\pi_{u_2}, \pi_{u_1}) = 0.7$, $Aff(\pi_{u_2}, \pi_{u_3}) = 0.6$ and $Aff(\pi_{u_2}, \pi_{u_4}) = 0.5$.

Other user-based predictions are computed: $\widehat{\pi}_{u_2}^{UCF}(i_4) = 0.722$, *and* $\widehat{\pi}_{u_2}^{UCF}(i_5) = 0.655$.

7.4.4 *Information Fusion*

One critical step in the Π HCF approach is to combine the predictions issued from user-based and item-based process. Actually, once we have computed the prediction on item i_n for user u_m separately according to user-based and item-based CF, the next step is to combine the two types of predictions into a single value (as depicted in Fig. 7.2). In the possibilistic framework, we will use the *adaptive fusion* [6] as a combination mode for the fusion of uncertain prediction information. This choice is justified by the fact that this mode allows a switch from conjunctive fusion to disjunctive one according to the amount of conflict $(Inc(\pi_{u_m}^{UCF}(i_n), \pi_{u_m}^{ICF}(i_n)))$ existing between the two sources S_1 (Item-based CF) and S_2 (User-based CF). Consequently, it chooses to combine three important criteria, namely the normalized conjunctive fusion, the disjunctive fusion and the degree of agreement. Intuitively, the norma lized conjunctive fusion makes sense when both sources (user-based and item-based CF) are considered as equally and fully reliable (i.e. They are consistent with each other). The disjunctive fusion should be applied when it is certain that at least one

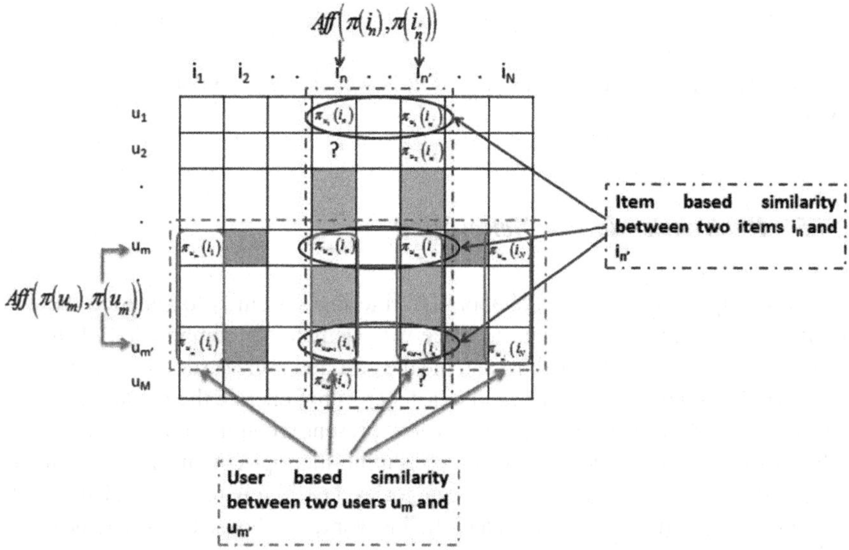

Fig. 7.2 Information fusion of the co-rated items and users

of the sources is reliable but it is not known which one. The degree of agreement $h(\pi_{u_m}^{UCF}(i_n), \pi_{u_m}^{ICF}(i_n))$ is used to normalize the results. The final prediction is given in Definition 2.

Definition 2 *The Information Fusion*: Let $\pi_{u_m}^{ICF}(i_n)$ and $\pi_{u_m}^{UCF}(i_n)$ be two possibilitic prediction degrees using respectively, item-based and user-based CF. The adaptive fusion, denoted by $\hat{\pi}_{u_m}^{AD}(i_n)$, is defined as follows:

$$\hat{\pi}_{u_m}^{AD}(i_n) = max(\hat{\pi}_{\wedge}(u_m(i_n)); min(\hat{\pi}_{\vee}(u_m(i_n)), 1 - h(\hat{\pi}_{u_m}^{UCF}(i_n), \hat{\pi}_{u_m}^{ICF}(i_n)))) \tag{7.10}$$

where:

- $\hat{\pi}_{\wedge}(u_m(i_n)) = \frac{min(\hat{\pi}_{u_m}^{UCF}(i_n), \hat{\pi}_{u_m}^{ICF}(i_n))}{h(\hat{\pi}_{u_m}^{UCF}(i_n), \hat{\pi}_{u_m}^{ICF}(i_n))}$ is the *normalized conjunctive fusion*.
- $\pi_{\vee}(u_m(i_n)) = max(\hat{\pi}_{u_m}^{UCF}(i_n), \hat{\pi}_{u_m}^{ICF}(i_n))$ is the *disjunctive operator*.
- $h(\hat{\pi}_{u_m}^{UCF}(i_n), \hat{\pi}_{u_m}^{ICF}(i_n)) = 1 - Inc(\hat{\pi}_{u_m}^{UCF}(i_n), \hat{\pi}_{u_m}^{ICF}(i_n)) = 1 - max(\hat{\pi}_{u_m}^{UCF}(i_n) \otimes \hat{\pi}_{u_m}^{ICF}(i_n))$ expresses the degree of agreement of the two sources.

Example 9 Since we have obtained the $\hat{\pi}_{u_2}^{UCF}(i_2)$, $\hat{\pi}_{u_2}^{UCF}(i_4)$ and $\hat{\pi}_{u_2}^{UCF}(i_5)$ predictions based on user-based CF and the $\hat{\pi}_{u_2}^{ICF}(i_2)$, $\hat{\pi}_{u_2}^{ICF}(i_4)$ and $\hat{\pi}_{u_2}^{ICF}(i)$ predictions based on item-based CF. Now, we combine results using the **adaptive fusion**, we obtain:
$\hat{\pi}_{u_2}^{AD}(i_2) = max(\pi_{\wedge}(u_2(i_2)); min(\pi_{\vee}(u_2(i_2)); 1 - h(\hat{\pi}_{u_2}^{UCF}(i_2), \hat{\pi}_{u_2}^{ICF}(i_2))))$
$\hat{\pi}_{u_2}^{AD}(i_2) = max(0.944; min(0.706; 0.294)) = \mathbf{0.944}$

where $h(\widehat{\pi}^{UCF}_{u_2}(i_2), \widehat{\pi}^{ICF}_{u_2}(i_2)) = 0.706$, $\pi_\wedge(u_2(i_2)) = \frac{0.667}{0.706} = 0.944$ and $\pi_\vee(u_2(i_2)) = 0.706$.

The others fusion predictions are as follows: $\widehat{\pi}^{AD}_{u_2}(i_4) = \mathbf{0.943}$ *and* $\widehat{\pi}^{AD}_{u_2}(i_5) = \mathbf{0.976}$.

7.4.5 Recommendation Generation

The most important step in a collaborative filtering system is to generate the output interface in terms of prediction. Once we calculate the combination of the co-rated items (item-based algorithm) and the the co-rated users (user-based algorithm) based on *the adaptive fusion mode*, the last step is to generate the recommendation list. In this context, there are various ways to present recommendations to the user either by offering the best items, or by presenting the top-K items as a recommendation list, or by classifying items into categories, i.e. 'highly recommended', 'fairly recommended' and 'not recommended'. In this work, we choose to present the top-K items as a recommendation list as it is the most used one [4, 8, 18].

Example 10 Once unknown ratings are predicted, the last step is to present the user u_2 top-**3** items as a recommendation list as follow:

$u_2, i_5, \widehat{\pi}^{AD}_{u_2}(i_5) = 0.976$
$u_2, i_2, \widehat{\pi}^{AD}_{u_2}(i_2) = 0.944$
$u_2, i_4, \widehat{\pi}^{AD}_{u_2}(i_4) = 0.943$

Item i_5 has the highest prediction value and consequently it will be in the top one recommendation list for user u_2 then item i_2 and finally i_4.

7.5 Experiments

In this section, we will describe the experimental protocol and we will present the experimental results of Π HCF.

7.5.1 Experimental Data

To evaluate the effectiveness of Π HCF, we will use the well-known movieLens data set available through the movieLens[2] website. This dataset is made up of a set of users preferences about movies. These preferences are votes that are integer values between 1 (dislike) and 5 (like). In our approach, as there are no uncertain datasets, we have chosen to convert these MovieLens preferences from [1,5] into a normalized

[2]1 http://movielens.org.

possibility degrees between [0.1], {0 (not satisfied) and 1 (fully satisfied)}. The data set contains in total 100.000 ratings collected by 943 users on 1682 movies, from 19-09-1997 to 22-04-1998. MovieLens data are represented as a sequence of events in the following way:

- user u_1 rates movie i_1 with 1,
- user u_1 rates movie i_3 with 0.5,
- user u_2 rates movie i_1 with 0.6, etc.

The dataset is divided into 2 parts, 80 % of the data is used for making predictions (the training set) and the 20 % left are the basis for measuring prediction accuracy (the test set). Each target user's ratings have been divided into a set of observed items and one of held-out items. The ratings of observed items are an input for predicting the ratings of held out items. To experimentally determine the impact of the test set size on the recommended movies quality, we propose to test three scenarios:

- **Set-1**: For each user, we randomly select 90 % of his ratings as instances in the training set and the remaining ones will be used in the testing set.
- **Set-2**: For each user, we randomly select 50 % of his ratings as instances in the training set and 10 % will be used in the testing set.
- **Set-3**: For each user, we randomly select 30 % of his ratings as instances in the training set and 10 % will be used in the testing set.

7.5.2 Evaluation Metrics

In order to evaluate the performance of recommender systems, several metrics have been proposed. According to [3], the evaluation metrics can be classified into three categories: *predictive accuracy metrics*, *classification accuracy metrics* and *rank accuracy metrics*. We introduce the commonly used CF metrics of each class.

1. **Predictive accuracy metrics**: measure how much the prediction p_i is close to the true numerical rating r_i expressed by the user. The evaluation can be done only for items that have been rated.

   - *Mean Absolute Error (MAE)* takes the mean of the absolute difference between each prediction and preference degree for all held-out preference degrees of users in the testing set. The lower the *MAE* the more accurately the recommendation engine predicts user ratings. Formally:

   $$MAE = \frac{\sum_{u_m,i_n} |\widehat{\pi}_{u_m}(i_n) - \pi_{u_m}(i_n)|}{N}, \tag{7.11}$$

   where N is the total number of preferences over all users, $\widehat{\pi}_{u_m}(i_n)$ is the predicted preference degree for user u_m on item i_n, and $\pi_{u_m}(i_n)$ is the actual preference.

Table 7.4 Confusion matrix

	Recommended	Not recommended
Relevant	RR	RN
Not relevant	FP	NN
Total	Rec = RR + FP	NRec = RN + NN

- *Normalized Mean Absolute Error (NMAE)* normalizes *MAE* to express errors as percentages:

$$NMAE = \frac{MAE}{\pi_{max} - \pi_{min}}, \tag{7.12}$$

where π_{max} and π_{min} are the upper and lower bounds of the preferences.

2. **Classification accuracy metrics**: evaluate how predictions help the active user in distinguishing good items from bad items. Therefore, it is useful in finding if the active user will like or not the current item. With classification metrics recommendation can be classified as:

   - *True positive (TP)*: an interesting item is recommended to the user.
   - *True negative (TN)*: an uninteresting item is not recommended to the user.
   - *False negative (FN)*: an interesting item is not recommended to the user.
   - *False positive (FP)*: an uninteresting item is recommended to the user.

   Precision and recall are the most popular metrics in the classification accuracy metrics. They are computed from a 2 × 2 table, such as the one shown in Table 7.4 where N is the number of items in the database.

   - *Precision:* is used to evaluate the validity of a given recommendation list. In fact, if an algorithm has a measured precision of 80 %, then the user can expect that, on average, 8 out of every 10 movies returned to the user will be used.

$$Precision = \frac{TP}{TP + FP} \tag{7.13}$$

   - *Recall:* computes the ratio of all used items that were recommended for active user relative to the total number of the objects actually collected.

$$Recall = \frac{TP}{TP + TN} \tag{7.14}$$

   - *F-measure:* combines both the precision and recall measures and indicates an overall utility of the recommendation list.

$$F - measure = \frac{2 * precision * recall}{precision + recall} \tag{7.15}$$

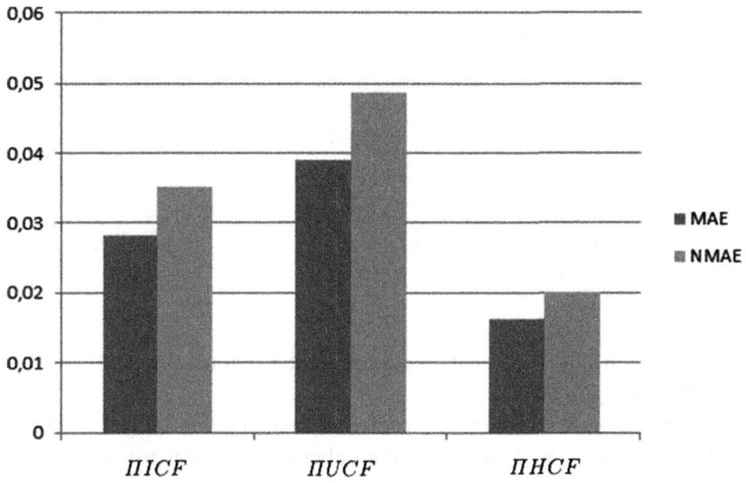

Fig. 7.3 The MAE and NMAE accuracy

In our experiment, we will adopt the most commonly used metrics [9], namely the *MAE* and *NMAE* measures for predictive accuracy and the *precision*, *recall* and *F-measure* for classification accuracy.

7.5.3 Experimental Results

In this section, we will first compare possibilistic approaches. Then, we will compare our possibilistic approach to the probabilistic one.

7.5.3.1 Possibilistic Comparison

In order to demonstrate the feasibility of the proposed approach *ΠHCF*, we will compare its results to the possibilistic item-based *ΠICF* and the possibilistic user-based *ΠUCF* using the *MAE* and *NMAE* as prediction measures and the precision, recall and F-measure as recommendation measures. The obtained results are summarized in Fig. 7.3 and Table 7.5.

1. *Prediction*: From Fig. 7.3, it is obvious that *Π* HCF performs better than *Π* ICF and *Π* UCF in terms of MAE and NMAE. In fact, we can deduce that the adaptive fusion computation has a clear advantage in prediction as the user-average error (equal to 0.0162) is significantly lower in this case. Similarly, the NMAE adaptive fusion measure has the lowest value (equal to 0.02) compared to the possibilistic user-based CF (equal to 0.0487) and the possibilistic item-based CF (equal to 0.0352). Therefore, we notice that combination contributes in the improvement

Table 7.5 The precision, recall and F-measure values for Set-1, Set-2 and Set-3

Approach	Set-1			Set-2			Set-3		
	Precision	Recall	F-measure	Precision	Recall	F-measure	Precision	Recall	F-measure
ΠICF (%)	39.71	48.57	43.69	35.78	48.57	41.2	28.4	31	29.64
ΠUCF (%)	30	35.04	32.32	27.3	33	29.88	23	30	26.03
ΠHCF (%)	**76.83**	**84.93**	**80.67**	**66.66**	**82.17**	**73.6**	**45.85**	**77.04**	**57.48**

of the prediction quality since it takes into consideration more predictions derived from different sources.

2. *Recommendation*: We will exploit preferences degrees in the whole recommendation list, Then, we will measure the number of items that are evaluated as relevant and/or irrelevant by the recommender. These measures are computed thrice using: our approach *Π* HCF, the possibilistic item-based *Π* ICF and the possibilistic user-based *Π* UCF. The obtained results are summarized in Table 7.5. We can notice that *Π* HCF approach outperforms *Π* ICF and *Π* UCF in terms of precision, recall and F-measure.
   In fact in set-3, our approach is able to provide a varied and good recommendation list even with a small set of users' data. Also, even with a large set of data *Π* HCF is still providing better recommendation. Likewise, the possibilistic hybrid CF method is affording a better recommendation accuracy (i.e. *Π* HCF proposes for one user an average of 80 good items out of every 100 recommended items while *Π* ICF offers an average of 40 good items out of every 100 recommended items and finally *Π* UCF gives an average of 30 good items per user). As a conclusion, we can say that *Π* HCF combination tends to increase the recommendation efficiency and yields to high recommendation performance. This confirms that combining *Π* ICF and *Π* UCF aims to solve the unavailability of data and to generate an efficient predictive model based on both users' and items' aspects.

7.5.3.2 Possibilistic Vs Probabilistic Approaches

In this sub-section, we will compare *Π*HCF to the standard probabilistic approach described in [24] and denoted by *PHCF*.

1. *Prediction*: Figure 7.4 summarizes MAE and NMAE results of *Π* HCF and *PHCF* approaches. In fact, *Π* HCF has a smaller MAE and NMAE than *PHCF*, meaning a better performance. This is explained by the fact that taking into consideration a purely uncertain user-item preferences matrix as input presents a significant improvement on prediction quality. Therefore, the possibilistic HCF outperforms the probabilistic one.

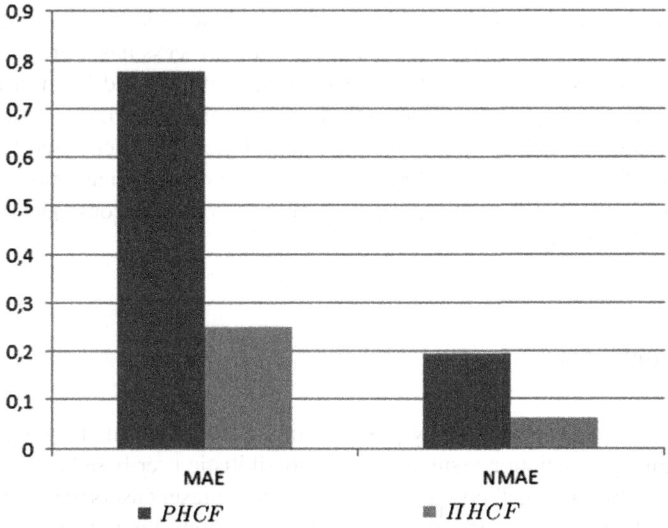

Fig. 7.4 The probabilistic versus possibilistic MAE and NMAE

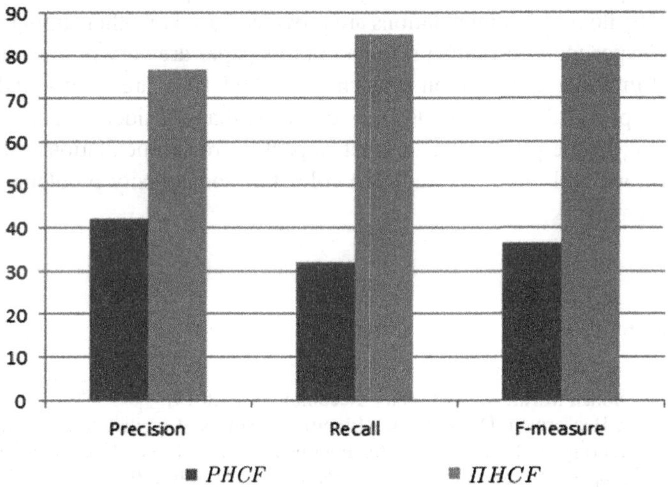

Fig. 7.5 Evaluation of classification accuracy metrics: Precision, recall and F-measure

2. *Recommendation*: From Fig. 7.5, results show that our approach outperforms *PHCF*. In fact, we pinpoint that the precision of our approach is equal to 76, 83 %, which is higher than that of *PHCF* (equal to 42, 1 %). Similarly, *Π* HCF's recall and F-measure are largely higher than *PHCF*'s ones.
   This explained by the fact that the *possibility theory*, especially the use of information affinity and the adaptive fusion measures have a considerable effect on

the recommendation quality rather than the absolute ratings. This confirms that considering uncertainty for the beginning of the process has a great impact on recommendation results. Furthermore, combining item-based and user-based CF methods has an appreciable effect in the improvement of recommendation results. As a result, combining on the one hand item-based and user-based CF and on the other hand the possibility theory framework have shown an improvement of the inter-connectivity accuracy among users and items and consequently a better quality of recommendation.

7.6 Conclusion

In this paper, we proposed a new possibilistic collaborative filtering approach Π HCF combining prediction results from the possibilistic user-based and item-based collaborative filtering methods. The basic idea is to represent users preferences using possibility theory framework, then compute predictions for both user-based and item-based approaches using the possibilistic *information affinity* measure. Next, we fused the two predictions lists into a single one using the possibilistic *adaptive fusion mode*. Finally, items recommendations are provided based on other similar users and items. The experimental results presented in this paper are very promising. In fact, we showed that recommendation performance of Π HCF are clearly higher compared to the probabilistic methods. This confirms that considering uncertainty for the beginning of the process has a great impact on recommendation results. As a future work, we will intend to study the cold start and sparsity problems under an uncertain framework.

References

1. Arekar, T., Sonar, M.R., Uke, N.J.: Surv. Recomm. Syst. (2014)
2. Breese, J.S., Heckerman, D., Kadie, C.: Empirical analysis of predictive algorithms for collaborative filtering. In: Proceedings of the Fourteenth Conference on Uncertainty in Artificial Intelligence, pp. 43–52. Morgan Kaufmann Publishers Inc, July 1998
3. Cremonesi, P., Turrin, R., Lentini, E., Matteucci, M.: An evaluation methodology for collaborative recommender systems. In: International Conference on Automated solutions for Cross Media Content and Multi-channel Distribution, 2008. AXMEDIS'08, pp. 224–231. IEEE, Nov 2008
4. Desrosiers, C., Karypis, G.: A comprehensive survey of neighborhood-based recommendation methods. In: Recommender Systems Handbook, pp. 107–144. Springer US (2011)
5. Dubois, D., Prade, H.: La fusion d'information imprécises. TS. Traitement du signal **11**(6), 447–458 (1994)
6. Dubois, D., Prade, H.: Possibility theory in information fusion. In: Proceedings of the Third International Conference on Information Fusion, 2000. FUSION 2000, vol. 1, pp. PS6–P19. IEEE, July 2000
7. Dubois, D., Prade, H.: Possibility theory and its applications: where do we stand?. In: Springer Handbook of Computational Intelligence, pp. 31–60. Springer, Berlin (2015)

8. Ekstrand, M.D., Riedl, J.T., Konstan, J.A.: Collaborative filtering recommender systems. Found. Trends Hum.-Comput. Interact. **4**(2), 81–173 (2011)
9. Herlocker, J.L., Konstan, J.A., Terveen, L.G., Riedl, J.T.: Evaluating collaborative filtering recommender systems. ACM Trans. Inf. Syst. (TOIS) **22**(1), 5–53 (2004)
10. Hu, R., Lu, Y.: A hybrid user and item-based collaborative filtering with smoothing on sparse data. In: 16th International Conference on Artificial Reality and Telexistence—Workshops, 2006. ICAT'06, pp. 184–189. IEEE, Nov 2006
11. Jenhani, I., Benferhat, S., Eloued, Z.: Possibilistic similarity measures. In: Foundations of Reasoning under Uncertainty, pp. 99–123. Springer, Berlin (2010)
12. Jin, R., Si, L., Zhai, C., Callan, J.: Collaborative filtering with decoupled models for preferences and ratings. In: Proceedings of the 12th International Conference on Information and Knowledge Management, pp. 309–316. ACM, Nov 2003
13. Kamishima, T.: Nantonac collaborative filtering: recommendation based on order responses. In: Proceedings of the 9th ACM SIGKDD International Conference on Knowledge Discovery and Data Mining, pp. 583–588. ACM, Aug 2003
14. Kantor, P. B., Rokach, L., Ricci, F., Shapira, B.: Recommender Systems Handbook. Springer (2011)
15. Liu, W.: Conflict analysis and merging operators selection in possibility theory. In: ECSQARU, pp. 816–827, Jan 2007
16. Miyahara, K., Pazzani, M.J.: Improvement of Collaborative Filtering with the Simple Bayesian Classifier 1 (2002)
17. Negoita, C., Zadeh, L., Zimmermann, H.: Fuzzy sets as a basis for a theory of possibility. Fuzzy Sets Syst. **1**, 3–28 (1978)
18. Sarwar, B., Karypis, G., Konstan, J., Riedl, J.: Item-based collaborative filtering recommendation algorithms. In: Proceedings of the 10th International Conference on World Wide Web, pp. 285–295. ACM, Apr 2001
19. Slokom, M., Ayachi, R.: A new item-based recommendation approach under a possibilistic framework. In: 5th International Workshop on Combinations of Intelligent Methods and Applications, CIMA (2015)
20. Smets, P., Kennes, R.: The transferable belief model. Artif. Intell. **66**(2), 191–234 (1994)
21. Su, X., Khoshgoftaar, T.M.: A survey of collaborative filtering techniques. Adv. Artif. Intell. **2009**, 4 (2009)
22. Tso-Sutter, K.H., Marinho, L.B., Schmidt-Thieme, L.: Tag-aware recommender systems by fusion of collaborative filtering algorithms. In: Proceedings of the 2008 ACM Symposium on Applied Computing, pp. 1995–1999. ACM, Mar 2008
23. Wang, J., Yin, J.: Combining user-based and item-based collaborative filtering techniques to improve recommendation diversity. In: 2013 6th International Conference on Biomedical Engineering and Informatics (BMEI), pp. 661–665. IEEE, Dec 2013
24. Wang, J., De Vries, A.P., Reinders, M.J.: Unifying user-based and item-based collaborative filtering approaches by similarity fusion. In: Proceedings of the 29th Annual International ACM SIGIR Conference on Research and Development in Information Retrieval, pp. 501–508. ACM, Aug 2006
25. Yamashita, A., Kawamura, H., Suzuki, K.: Adaptive fusion method for user-based and item-based collaborative filtering. Adv. Complex Syst. **14**(02), 133–149 (2011)

Printed in the United States
By Bookmasters